Introduction to WCDMA

Physical Channels, Logical Channels, Network, and Operation

Lawrence Harte

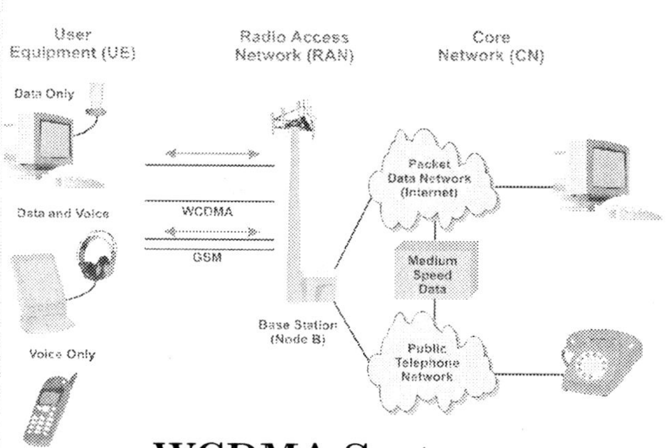

WCDMA System

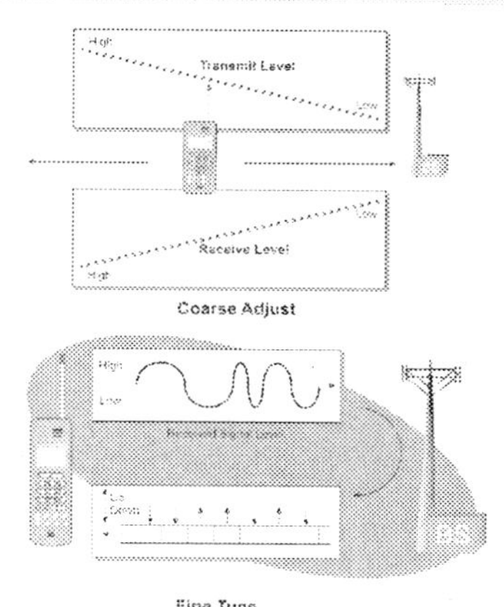

RF Power Control

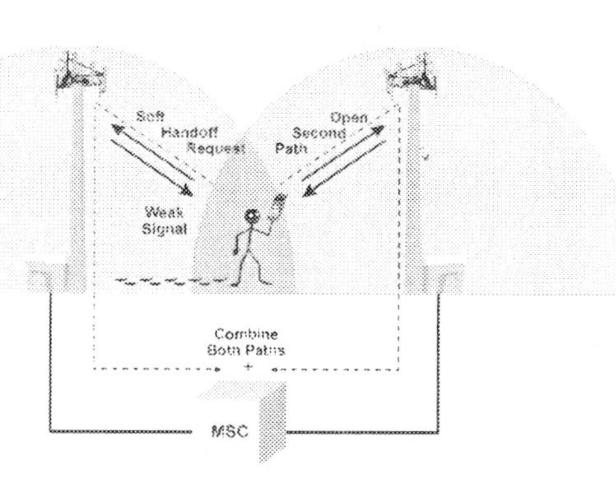

Soft Handover

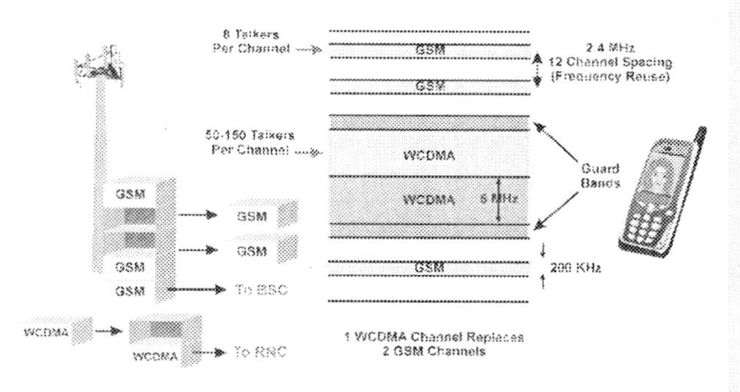

Upgrading GSM to CDMA

Excerpted From:

Wireless Systems

With Updated Information

ALTHOS Publishing

ALTHOS Publishing

ISBN: 1-9328131-2-8

ALTHOS electronic books (ebooks) and images are available for use in educational, promotional materials, training programs, and other uses. For more information about using ALTHOS ebooks and images, please contact Karen Bunn at kbunn@Althos.com or (919) 557-2260

About the Authors

Mr. Harte is the president of Althos, an expert information provider which researches, trains, and publishes on technology and business industries. He has over 29 years of technology analysis, development, implementation, and business management experience. Mr. Harte has worked for leading companies including Ericsson/General Electric, Audiovox/Toshiba and Westinghouse and has consulted for hundreds of other companies. Mr. Harte continually researches, analyzes, and tests new communication technologies, applications, and services. He has authored over 50 books on telecommunications technologies and business systems covering topics such as mobile telephone systems, data communications, voice over data networks, broadband, prepaid services, billing systems, sales, and Internet marketing. Mr. Harte holds many degrees and certificates including an Executive MBA from Wake Forest University (1995) and a BSET from the University of the State of New York, (1990).

Table of Contents

Introduction to WCDMA

Wideband Code Division Multiple Access (WCDMA)

The 3rd generation wideband code division multiple access (WCDMA) system is a mobile radio communication system that provides for high-speed data and voice communication services. WCDMA is one of two technologies that are being used to fulfill the radio access requirements of universal mobile telecommunications system (UMTS).

Installing a new WCDMA system or upgrading existing systems to WCDMA allows mobile service providers to offer their customers wireless broadband (high-speed Internet) services and to operate their systems more efficiently (more customers per cell site radio tower).

The WCDMA system is composed of mobile devices (wireless telephones and data communication devices called user equipment - UE), radio towers (cell sites called Node Bs), and an interconnection system (switches and data routers). The WCDMA system uses two types of radio channels; frequency division duplex (FDD) and time division duplex (TDD). The FDD radio channels are primarily used for wide area voice (audio) channels and data services. The TDD channels are typically used for systems that do not have the availability of dual frequency bands.

The WCDMA system has been designed to interoperate with GSM systems. This allows for the gradual migration of GSM customers to advanced WCDMA digital services.

WCDMA radio towers (cell sites) are composed of an antenna system, and radio equipment (base station). WCDMA base stations contain one or more WCDMA digital and/or GSM radio channels. The base station converts radio signals from mobile devices into electrical signals that can be transferred to the cellular interconnection system (typically a mobile switching center). In a typical cellular system, each wide WCDMA radio channel typically replaces two GSM radio channels.

Cell sites are typically interconnected to each other through a central switching system called a mobile switching center (MSC) and or packet data routers. The MSC is a circuit switching systems as it allows continuous communication between a mobile device and cell sites. The packet data servicing node (PDSN) is a packet router that receives and forwards packets towards their destination.

Figure 1.1 shows a simplified diagram of a WCDMA system. This diagram shows that the WCDMA system includes various types of mobile communication devices (called user equipment - UE) that communicate through base stations (node B) and a mobile switching center (MSC) or data routing networks to connect to other mobile telephones, public telephones, or to the Internet via a core network (CN). This diagram shows that the WCDMA system is compatible with both the 5 MHz wide WCDMA radio channel and the narrow 200 kHz GSM channels. This example also shows that the core network is essentially divided between voice systems (circuit switching) and packet data (packet switching).

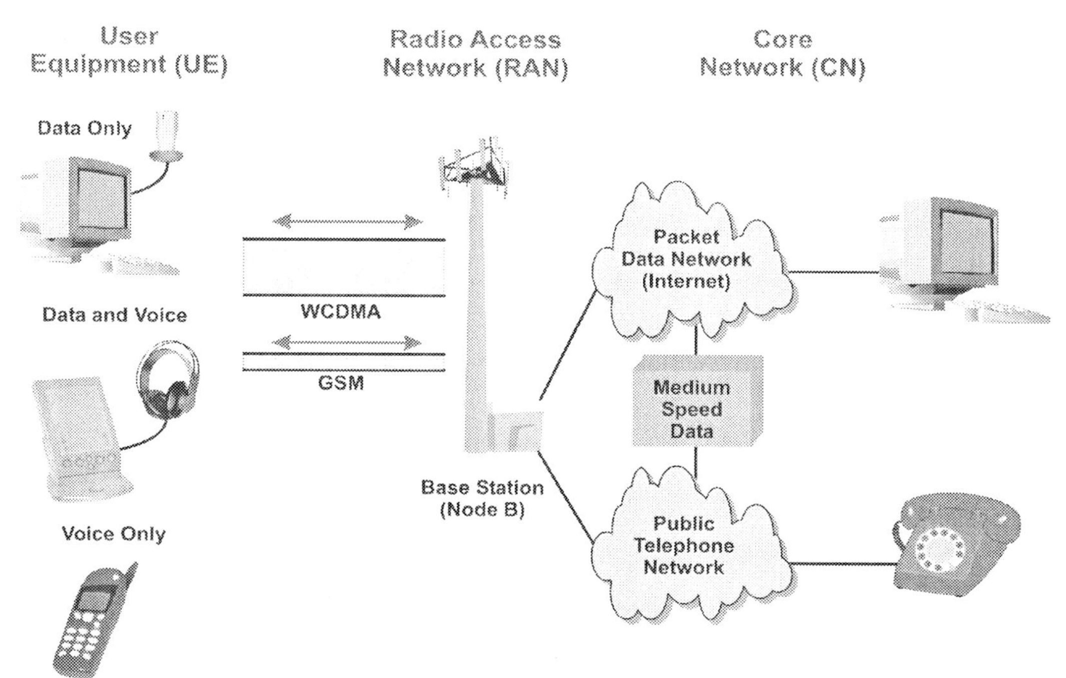

Figure 1.1., WCDMA System Overview

The WCDMA system allows cellular carriers to offer new broadband wireless Internet services for existing and new customers. The WCDMA system is designed to permit advanced and reliable services including media streaming and large file transfers. These new services offer the potential of higher average revenue per user (ARPU) than existing 1st and 2nd generation mobile customers. For existing mobile carriers that upgrade to WCDMA (such as GSM carriers), WCDMA capability allows the carriers to market to new data-only customers they don't already have.

Customers can access the high-speed Internet services through WCDMA capable handsets or external modems that connect to their desktop or laptop computers. The WCDMA radio channels are an "always-on" system that allows users to browse the Internet without complicated dialup connections. Because the WCDMA system was designed to provide multimedia services, it is plug-and-play with most computer operating systems.

Each 5 MHz WCDMA RF channel can have more than 100 separate coded communication channels. Some of the channels are used for control purposes and the remaining channels are used for voice (audio) and user data transmission. The number of coded channels can vary based on the spreading factor used. 3rd generation WCDMA systems are more efficient than 2nd generation systems being that they allow 50 to 150 mobile devices to simultaneously share a single radio channel for voice and data communication.

The key attributes of a WCDMA system include a wide 5 MHz bandwidth CDMA radio channel, the co-existence of multiple physical channels on the same frequency using channel codes, many logical (transport) channels, multiple signaling methods, increased frequency reuse, multiple speech coding technologies, improved paging methods, muli-system operation, and other advanced operational features.

The system components of the WCDMA system are the same basic types of subsystems as the GSM system. Some of the names of the network parts have been changed and new functions have been added. WCDMA systems are composed of three basic parts; mobile devices (called user equipment – UE), radio network (called UMTS terrestrial access network -UTRAN), and an interconnection network (called the core network – CN).

WCDMA System Evolution

The WCDMA system has evolved from 1st generation analog systems (primarily voice), through 2nd generation digital (voice and low speed data), through evolved 2nd generation GPRS and EDGE digital (medium speed packet data), to provide multimedia broadband capability.

In the 1990s, it became clear that 2nd generation digital systems could not meet the growing needs for mobile communication. In 1992, the world administrative radio conference (WARC) defined new frequency bands that could provide new 3rd generation mobile services. The initial requirements for 3rd generation systems were defined in an international mobile telecommunications 2000 (IMT-2000). The IMT-2000 key requirements included multimedia communications, data transmission rates of greater than 2 Mbps, and global mobility. IMT-2000 is managed by the international telecommunications union (ITU).

Using the IMT-2000 requirements, a universal mobile telephone service (UMTS) was created. The UMTS system defines a complete system for providing universal communication including mobile and fixed based communication. For land based mobile communication it is composed of a UMTS terrestrial radio access network (UTRAN), a core network (CN), and gateways that connect the core network to other systems (such as the public telephone network and the Internet). You can find out more information about UMTS at www.UMTS-Forum.org.

One of the radio access technologies used for UTRAN is the wideband code division multiple access (WCDMA) system (as of 2004 there were 2 access technologies used). The WCDMA technical specification was initially developed to allow a single global mobile communication system.

Initially, the WCDMA system was planned to be a completely new network, radio system and infrastructure. Because the development of new infrastructure equipment is expensive and time consuming and many carriers have already invested in GSM infrastructure, the WCDMA system primarily defines a new radio access network (RAN) and the infrastructure (switching systems) remains the same as the GSM system. The industry specification development effort began in 1996 and the first commercial WCDMA mobile system was launched in Japan in 2001.

3rd Generation Partnership Project (3GPP)

The 3GPP group is a collaborative organization that works on the creation of 3rd generation industry global standards that provide for high-speed multimedia wireless services. The 3rd generation partnership project (3GPP) oversees the creation of industry standards for the 3rd generation of mobile wireless communication systems (WCDMA). The key members of the 3GPP include standards agencies from Japan, Europe, Korea, China and the United States.

WCDMA Industry Specifications

WCDMA is a group of specifications that define the radio part of the 3rd Generation UMTS wireless systems. The WCDMA system is specified by the 3rd generation partnership project (3GPP) and you can obtain the WCDMA and related specifications at www.3GPP.org.

Upgrading GSM to WCDMA

The deployment of WCDMA allows carriers to upgrade existing GSM systems with new software and some new hardware such as replacing or adding radio transceivers and adding new radio network controllers (RNCs). Service providers (operators) can upgrade their GSM systems to offer WCDMA services by either adding WCDMA channels and/or replacing existing GSM channels.

The WCDMA system was designed to simplify the requirements of the node B radio equipment by moving some of the radio control functions to the radio network controller (RNC). As a result, the RNC for the WCDMA system is more sophisticated than the base station controller (BSC) for the GSM system.

While WCDMA radio channels can operate on the same frequency in adjacent cell sites, GSM radio channels typically require frequency planning that requires them to be spaced 12 channels apart. This means that for each 5 MHz CDMA channel, two or more 200 kHz radio channels must be removed or not used. In addition, WCDMA radio channels require a guard band (a radio buffer zone) between other radio channels. This guard band slightly increases the number of radio channels that must be removed or not used in a cell site.

Figure 1.2 shows how a GSM system can be upgraded to offer WCDMA services. This diagram shows that typically 2 or more GSM channels are typically removed to allow a new WCDMA channel to be added. This example also shows that a separate base station assembly is added to communicate with a WCDMA RNC as the GSM BSC is not equipped to manage the WCDMA radio channels.

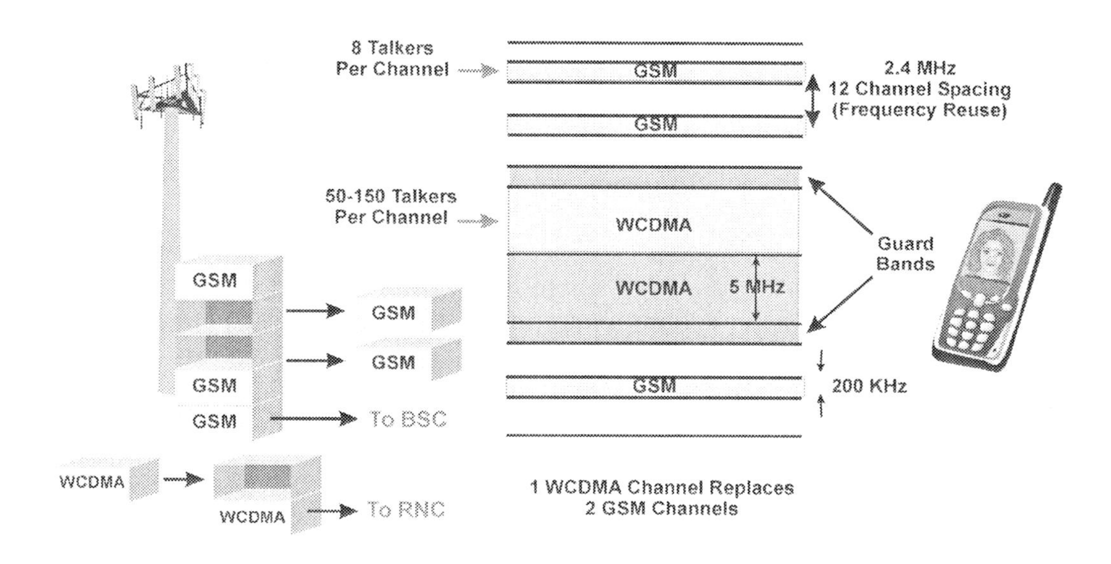

Figure 1.2., GSM System Upgrade to WCDMA

Because each WCDMA radio channel can simultaneously provide audio connections to over 100 voice and data customers and each GSM radio channel can only provide 8 audio connections per radio channel, This means that the installation of WCDMA radio channels typically increases the number of customers that can be served in the same frequency range by 5 to 10 times the existing GSM customers.

Dual Mode Capability

Dual mode (multi-mode operation) capability is the ability of a device or system to operate in two different modes (not necessarily at the same time). For wireless systems, it refers to mobile devices that can operate on two different system types such as analog and digital or different forms of digital.

The WCDMA system was designed to allow calls to transfer to and from existing GSM systems to WCDMA systems. This allows for the gradual transfer of customers from 2nd generation GSM systems to 3rd generation WCDMA systems.

Figure 1.3 shows how the WCDMA system can interoperate with either the WCDMA system or GSM system. This diagram shows how a mobile device is operating between a GSM and a WCDMA system. This example shows that the WCDMA system and GSM system share the same network systems so call completion and transfer (handoff) can be transparent to the user.

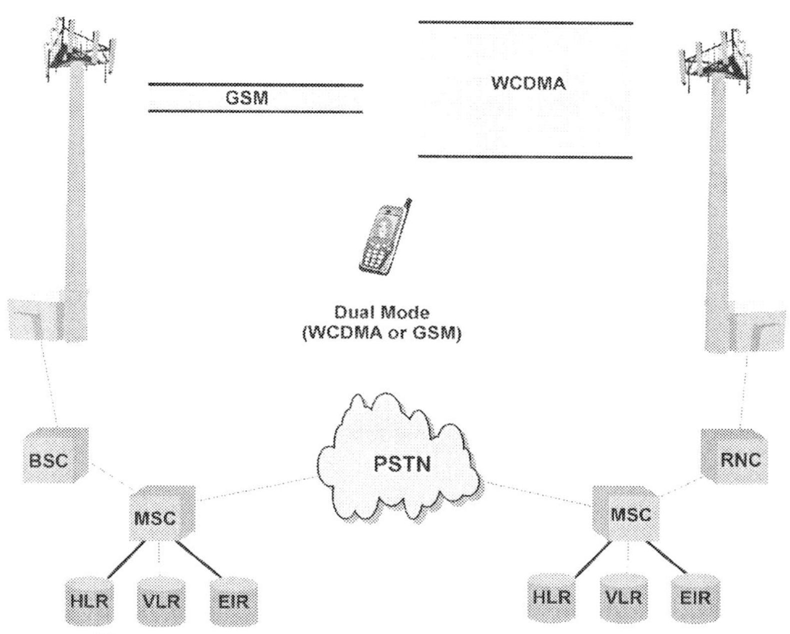

Figure 1.3., WCDMA and IS-95 Dual Mode Operation

WCDMA Services

The services that WCDMA can provide include voice services, data services, multicast services, location based services (LBS), and multimedia communication services that have various levels of quality of service (QoS).

Voice Services

Voice service is a type of communication service where two or more people can transfer information in the voice frequency band (not necessarily voice signals) through a communication network. Voice service involves the setup of communication sessions between two (or more) users that allows for the real time (or near real time) transfer of voice type signals between users.

The WCDMA system provides for various types of digital voice services. The voice service quality on the WCDMA system can vary based on a variety of factors. The WCDMA system can dynamically change the voice quality because the WCDMA system can use several different types of speech compression. The service provider can select and control which speech compression process (voice coding) is used. The selection of voice coders that have higher levels of speech compression (higher compression results in less digital bits transmitted) allows the service provider to increase the number of customers it can provide service to with the tradeoff of providing lower quality audio signals.

Data Services

Data Services are communication services that transfer information between two or more devices. Data services may be provided in or outside the audio frequency band through a communication network. Data service involves the establishment of physical and logical communication sessions between two (or more) users that allows for the non-real time or near-real time transfer of data (binary) type signals between users.

When data signals are transmitted on a non-digital channel (such as an analog telephone line), a data modem must be used. The data modem converts the data signal (digital bits) into tones that can be transferred in the audio frequency band. Because the speech coder used in the WCDMA system only compresses voice signals and not data modem signals, analog modem data cannot be sent on a WCDMA traffic (voice) channel.

When data signals are transmitted on a WCDMA radio channel, a data transfer adapter (DTA) is used. The DTA converts the data bits from a computing device into a format that is suitable for transmission on a communication channel that has a different data transmission format. DTAs are used to connect communication devices (such as a PDA or laptop) to a mobile device when it is operating on a WCDMA digital radio channel.

The data service that WCDMA radio channels can provide include low-speed packet data up to more than 2 Mbps. Each individual WCDMA traffic channel (single logical channel on a radio channel) is capable of using variable channel spreading rates to provide for high speed data transmission. To get even higher speed data channels, multiple traffic channels (up to 4 or 6) can be combined.

Bandwidth on demand (BOD) is a process that allows the data transmission rate for a specific user to dynamically change based on requests from the customer, their application (e.g. voice or video), and the data transmission capability of the system.

Circuit Switched Data

Circuit switched data is a data communication method that maintains a dedicated communications path between two communication devices regardless of the amount of data that is sent between the devices. This gives to communications equipment the exclusive use of the circuit that connects them, even when the circuit is momentarily idle.

To establish a circuit-switched data connection, the address is sent first and a connection (possibly a virtual non-physical connection) path is established. After this path is setup, data is continually transferred using this path until the path is disconnected by request from the sender or receiver of data.

Packet Switched Data

Packet switch data is the transfer of information between two points through the division of the data into small packets. The packets are routed (switched) through the network and reconnected at the other end to recreate the original data. Each data packet contains the address of its destination. This allows each packet to take a different route through the network to reach its destination.

WCDMA packet-switched data service is an "always-on" type of service. When the WCDMA device is initially turned on, it takes only a few seconds to obtain an IP address that is necessary to communicate with the network. Even when the WCDMA device is inactive and placed in the dormant state, reconnection is typically less than 1/2 a second.

Multicast Services

Multicast service is a one-to-many media delivery process that sends a single message or information transmission that contains an address (code) that is designated for several devices (nodes) in a network. Devices must contain the matching code to successfully receive or decode the message. WCDMA multicast services can include news services or media (digital audio) broadcasts.

Location Based Services (LBS)

Location based services are information or advertising services that vary based on the location of the user. The WCDMA system permits the use of different types of location information sources including the system itself or through the use of a global positioning system (GPS).

Messaging Services

Messaging services are the transfer of short information messages between two or more users in a communication system. The WCDMA system allows for mobile messaging services that can transfer up to 160 characters per message.

Quality of Service (QoS)

Quality of service (QoS) is one or more measurements of desired performance and priorities of a communications system. QoS measures may include service availability, maximum bit error rate (BER), minimum committed bit rate (CBR) and other measurements that are used to ensure quality communications service.

The WCDMA system can offer different types of quality of service (QoS) for different types of customers and their applications. A key QoS attribute includes priority access for different types of users. For example, real time priority access typically applies to voice services and reliable data transfer is applied to interactive data services.

Conversation Class

Conversation class is the providing of communication service (typically voice) through a network with minimal delay in two directions. While conversation has stringent maximum time delay limits (typically tens of milliseconds), it is typically acceptable to loose some data during transmission due to errors or discarding packets during system overcapacity.

Streaming Class

Streaming class is the delivering of audio or video signals through a network by establishing and managing of a continuous flow (a stream) of information. Upon request of streaming class of service, a server system (information source) will deliver a stream of audio and/or video (usually compressed) to a client. The client will receive the data stream and (after a short buffering delay) decode the audio and play it to a user. Internet audio streaming systems are used for delivering audio from 2 kbps (for low-quality speech) up to hundreds of kbps (for audiophile-quality music).

Streaming class provides a continuous stream of information that is commonly used for the delivery of audio and video content with minimal delay (e.g. real-time). Streaming signals are usually compressed and error protected to allow the receiver to buffer, decompress, and time sequence information before it is displayed in its original format.

Interactive Class

Interactive class is the providing of data and control information through a network with minimal delays and with very few data errors. Interactive class allows a user or system to interact with a software application (typically a web host) in near-real time (limited transmission delays). Interactive class allows the communication channel to be shared by other users during periods of inactivity (such as when the user is thinking about a response to a web page question.)

Background Class

Background class is the process of providing information transfer services on a best-effort basis. Background class is used for non-time critical services (such as Internet web browsing).

WCDMA Products (Mobile Devices)

WCDMA mobile devices (also called access terminals) are data input and output devices that are used to communicate with a nearby access point. WCDMA devices include removable UMTS subscriber identity modules (USIMs) that hold service subscription information. The common types of available WCDMA devices include external radio modems, PCMCIA cards, radio modules, and dual mode mobile telephones.

UMTS Subscriber Identity Module (USIM)

A UMTS subscriber identity module (USIM) is an "information" card that contains service subscription identity and personal information. This information includes a phone number, billing identification information and a small amount of user specific data (such as feature preferences and short messages.) This information can be stored in the card rather than programming this information into the phone itself. This intelligent card, either credit card-sized (ISO format), or the size of a postage-stamp (Plug-In format), can be inserted into any UIM ready wireless device.

PCMCIA Air Cards

The PCMCIA card uses a standard physical and electrical interface that is used to connect memory and communication devices to computers, typically laptops. The physical card sizes are similar to the size of a credit card 2.126 inches (51.46 mm) by 3.37 inches (69.2 mm) long. There are 4 different card thickness dimensions: 3.3 (type 1), 5.0 (type 2), 10.5 (type 3), and 16 mm (type 4). WCDMA PCMCIA radio cards can be added to most laptop computers to avoid the need of integrating or attaching radio devices.

Embedded Radio Modules

Embedded radio modules are self contained electronic assemblies that may be inserted or attached to other electronic devices or systems. Embedded radio modules may be installed in computing devices such as personal digital assistants (PDAs), laptop computers, and other types of computing devices that can benefit from wireless data and/or voice connections.

Mobile Telephones

Mobile telephones are radio transceivers (combined transmitter and receive) that convert signals between users (typically people, but not always) and radio signals. Mobile telephones can vary from simple voice units to advanced multimedia personal digital assistants (PDAs). WCDMA mobile telephones may include both GSM and WCDMA capability (dual mode). This allows dual mode WCDMA mobile device to access both 2nd generation GSM systems and 3rd generation WCDMA systems.

External Radio Modems

External radio modems are self contained radios with data modems that allow the customer to simply plug the radio device to their USB or Ethernet data port on their desktop or laptop computer. External modems are commonly connected to computers via standard connections such as universal serial bus (USB) or RJ-45 Ethernet connections.

Figure 1.4 shows the common types of WCDMA products available to customers. This diagram shows that the product types available for WCDMA include dual mode and single mode mobile telephones, PCMCIA data cards, integrated (embedded) radio modules, and external radio modems. WCDMA mobile telephones may be capable of operating on both the WCDMA voice and analog radio channels or on the data only WCDMA channels "dual mode". PCMCIA data cards may allow for both data and voice operations when inserted into portable communications devices such as laptops or personal digital assistants (PDAs). Small radio assemblies may be integrated (embedded) into other devices such as laptop computers or custom communication devices. External modems may be used to provide data services to fixed users (such as desktop computers).

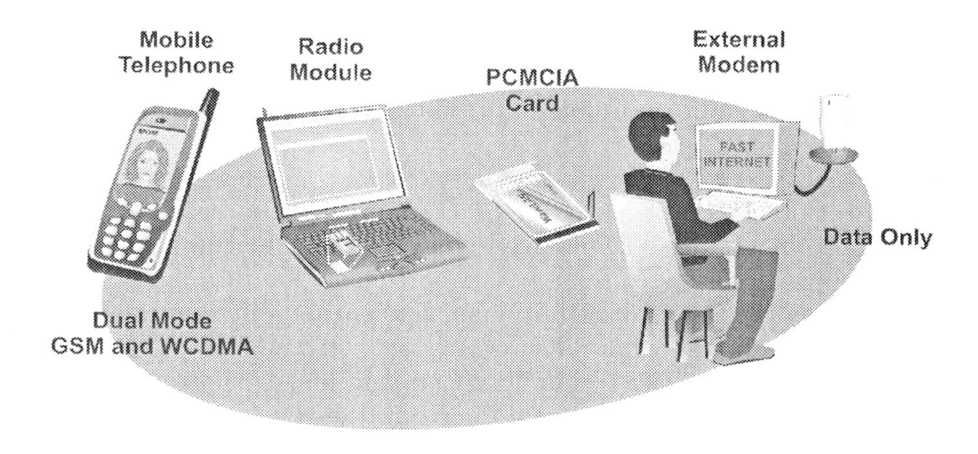

Figure 1.4., WCDMA Product Types

WCDMA Radio

There are two types of radio systems used by the WCDMA system; frequency division duplex (FDD) and time division duplex (TDD). The characteristics of WCDMA radio channels include frequency bands, frequency reuse, channel multiplexing techniques, RF Power levels, RF power control, and channel structure.

WCDMA radio channels are divided into overlapping physical (transport) and logical channels. Each physical radio channel is uniquely identified by a spreading code. The logical channels (e.g. control, voice and data) are divided into groups of bits (fields and frames).

Frequency Division Duplex (FDD) WCDMA

Frequency division duplex (FDD) is a communications channel that allows the transmission of information in both directions via separate bands (frequency division). The WCDMA frequency division duplex (FDD) system pairs two 5 MHz radio channels to provide for simultaneous duplex communication.

FDD operation normally assigns the transmitter and receiver to different communication channels. One frequency is used to communicate in one direction and the other frequency is required to communicate in the opposite direction. The amount of frequency separation between the transmitter and receiver frequency bands varies based on the range of frequencies. Generally, the higher the frequency used, the larger the amount of duplex frequency spacing. The lower frequency band of the pair is generally designated for the mobile because it is easier (and often cheaper) to produce lower frequency devices.

Each radio channel is divided into 10 msec frames and each frame is further divided into 15 time slots (666 usec each). During a typical FDD voice conversation, all the time slots are dedicated for transmitting and for receiving.

Because there are times that a mobile device must perform other radio signal processing functions on other frequencies during conversation (during voice transmission), the mobile device may enter into a compressed mode of transmission. Compressed mode allows the mobile device to compress a 10 msec frame of data into only some of the 15 time slots within the 10 msec data frame (sending a higher data rate in some of the other slots). This allows the mobile device to use the time during the unused time slots to perform other functions such as retuning and measuring the signal strength and quality of other radio channels.

Figure 1.5 shows the basic parts and operation of the WCDMA FDD system. This diagram shows that the WCDMA FDD system uses two 5 MHz radio channels that are nominally separated by 190 MHz (85 MHz separation in USA) to provide for full duplex two-way communication. Each radio channel is further divided into 10 msec frames and 15 time slots (666 usec each) that contain control and user data.

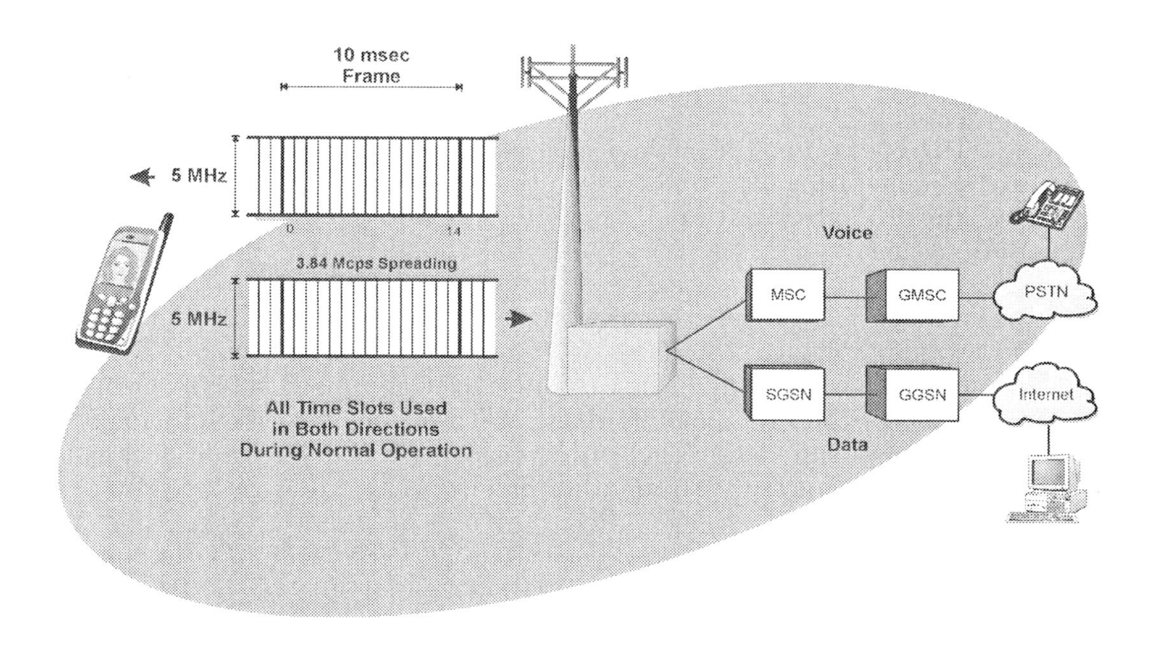

Figure 1.5., Frequency Division Duplex (FDD) WCDMA System

Time Division Duplex (TDD) WCDMA

Time division duplex (TDD) is a process of allowing two way communications between two devices by time sharing. When using TDD, one device transmits (device 1), the other device listens (device 2) for a short period of time. After the transmission is complete, the devices reverse their role so device 1 becomes a receiver and device 2 becomes a transmitter. The process continually repeats itself so data appears to flow in both directions simultaneously.

The WCDMA system can use time division duplex (TDD) format technology when paired frequencies are not available. The WCDMA TDD system uses the same CDMA channel coding process as the WCDMA FDD system. The WCDMA TDD system is well suited for indoor areas and wide area data systems.

During TDD operation, the mobile device receives a burst of data, waits until its assigned time slot, and transmits a burst of data on the same frequency. This process is continually repeated providing a steady flow of data in both directions.

A key challenge of TDD operation is the potential that the transmission time from mobile devices varies with the distance that the mobile device operates from the cell site and this can cause overlapping with transmission bursts received from other mobile devices. To overcome this challenge, a guard time period is added to each transmission burst to ensure it does not overlap with transmit bursts from other mobiles.

Figure 1.6 shows the basic parts and operation of the WCDMA TDD system. This diagram shows that the WCDMA TDD system divides the 5 MHz radio channel 10 msec frames into 15 time slots (666 usec each). Time slot can be transmitted in either the uplink or downlink direction. This example shows that time slots can be dynamically assigned so the system can provide variable and asymmetrical data transmission rates.

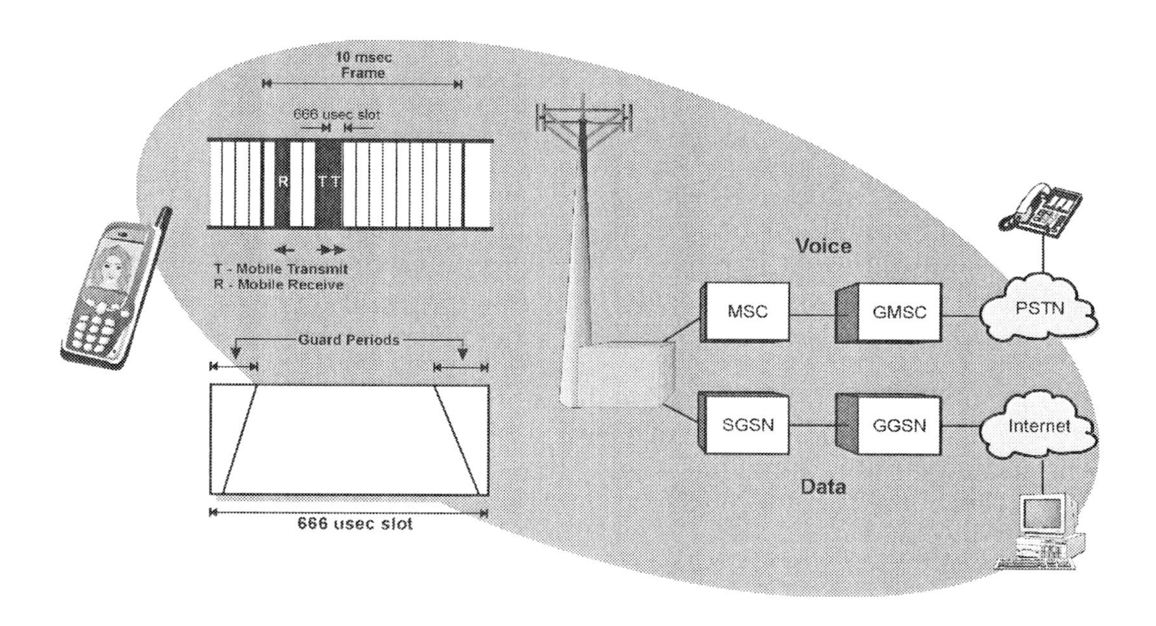

Figure 1.6., Time Division Duplex (TDD) WCDMA - TDD/WCDMA System

RF Channel Multiplexing

RF channel multiplexing is a process that divides a single radio transmission path into several parts that can transfer multiple communication (voice and/or data) channels. Multiplexing may be frequency division (dividing into frequency bands), time division (dividing into time slots), code division (dividing into coded data that randomly overlap), or statistical multiplexing (dynamically assigning portions of channels when activity exists).

Code division multiplexing (CDM) is a system that allows multiple users to share one or more radio channels for service by adding unique codes to each data signal that is being sent to and from each of the radio transceivers. These codes are used to spread the data signal to a bandwidth much wider than is necessary to transmit the data signal without the code.

Multiple coded (physical) channels can be used on the same RF channel without interfering with each other by mixing in channelization codes that uniquely identify each physical channel. This is because the channelisation codes used are pre-selected to be orthogonal (non-interfering) with each other.

RF channel spreading is the process of creating a radio communication channel that occupies a radio channel bandwidth that is larger than frequency bandwidth that is required to only send the information signal. To create larger RF channels, the information may be multiplied by additional signals (channel codes).

Variable spreading rates are the ability of the ratio of chips to information bits (spreading rate) to change in a spread spectrum communication system. The use of variable spreading rates allows the system to assign different data transmission rates on the same radio channel without having to change the chip rate.

Channel Codes

Channel codes are unique patterns or codes that are combined with, mixed into, or used to modify information that is sent on a communication channel to identify each channel that is sent on a common transmission channel. The WCDMA system uses channel codes to expand information signals. Different channel codes are used to identify different physical channels.

Each bit of information that is to be transmitted in the WCDMA system is multiplied (spread) by a channel code. This produces a relatively long sequence of information elements (called chips) that represent each bit of digital information. The number of code chips that is transmitted on a radio channel is fixed at 3.84 million chips per second (Mcps).

The ratio of how many chips are used to represent each bit of information that is transmitted is known at the spreading factor (SF). Each radio channel can transfer multiple coded channels that have different spreading factors. The spreading factors on the WCDMA system vary from 4 to 512 on the downlink channel and from 4 to 256 on the uplink channel.

Because several chips of information represent each bit of information, even with a loss of a few chips due to collisions with other signals still allows a majority of the chips to be used to correctly recreate the original information bits.

In general, the use of a higher spreading factor provides for a more robust (reliable) communication channel as more chips represent each information bit. However, using a higher amount of spreading results in a lower data transmission rate available to the user. The WCDMA varies the spreading factor (number of chips per bit) to provide bandwidth on demand (BoD) service.

Figure 1.7 shows an example of some of the spreading rates for the downlink channel. This table shows that the WCDMA system has a gross (unprotected) channel data transfer rate that ranges from 15 kbps to 1920 kbps depending on the spreading factor. This table also shows that if ½ rate convolutional coding is used (it does not have to be used), the net data rate is less than half of the gross channel data rate due to the use of error protection and system signaling overhead. This example shows that this results in a net data transmission rate that varies from 3 kbps to 936 kbps for each coded radio communication channel.

Spreading Factor (SF)	Gross Channel Data Rate (kbps)	Net Data Rate (kbps) (with 1/2 rate convolutional coding + overhead)
4 (minimum spreading)	1920	936
8	960	456
16	480	215
32	240	105
64	120	45
128	60	12
256	30	6
512	15	3

Figure 1.7., Downlink Radio Channel Spreading Table

Figure 1.8 shows the spreading rates for the uplink channel. This table shows that the WCDMA system has a gross (unprotected) channel data transfer rate that ranges from 15 kbps to 960 kbps depending on the spreading factor. This table also shows that the net data rate is less than half of the gross channel data rate due to the use of ½ rate convolutional coding. This results in a net data transmission rate that varies from 7.5 kbps to 480 kbps for each coded radio communication channel.

Spreading Factor (SF)	Gross Channel Data Rate (kbps)	Net Data Rate (kbps) (with 1/2 rate convolutional coding)
4 (minimum spreading)	960	480
8	480	240
16	240	120
32	120	60
64	60	30
128	30	15
256	15	7.5

Figure 1.8., Uplink Radio Channel Spreading

Parallel Channel Codes

Parallel channel codes are combined coded channels that are transmitted on the same transmission channel. Parallel channel codes are used in the WCDMA to increase the overall data transmission throughput on a single radio channel.

The WCDMA system can combine up to 4 channels in the forward (downlink) direction and up to 6 channels in the reverse (uplink) direction to provide for data transmission rates above 2 Mbps.

To coordinate parallel channels codes, the system coordinates the parallel channels by transmitting a control message that identifies how the independent physical channels and their associated logical channels are combined or divided. A transport format combination indicator (TFCI) code is used to coordinate the combination of physical channels and transport format indicator (TFI) coordinates each transport channel. To use parallel channel codes, the channel must use the minimum spreading factor (SF).

Figure 1.9 shows how the WCDMA can combine multiple coded physical channels to provide higher data transmission rates than is possible with a single coded channel. This diagram shows that the same frequency bandwidth can be used to simultaneously transmit several 5 MHz coded channels. Each RF carrier channel is differentiated by a different scrambling code. This example shows that the combining of these channels can produce broadband data transmission rates.

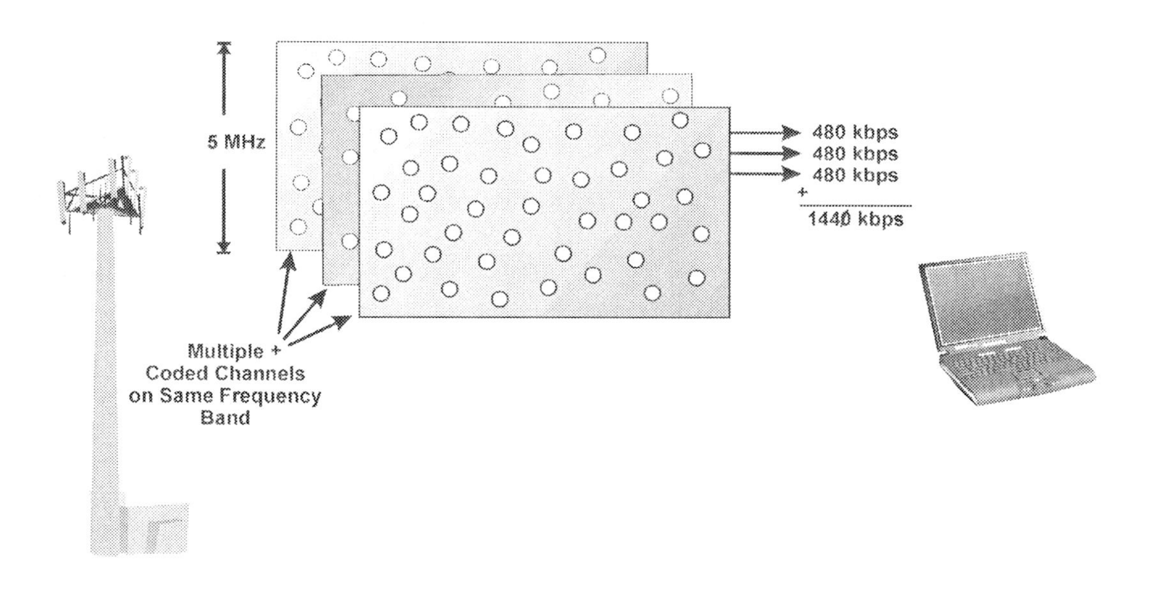

Figure 1.9., Combining Coded Channels

Channel Code Trees

Channel code trees are hierarchical structured code sequences that allow codes to be selected from a table that do not interfere with each other. By structuring the codes in a tree format, lower level codes (branches of the tree) are subsets of upper level codes. When lower level codes are selected, the upper level codes cannot be used.

Figure 1.10 shows how the WCDMA system uses a code tree to assign different data rates to different users. This example shows that upper level codes in the code tree can be used to provide higher data rates and that these upper level codes use a lower spreading factor that provides for the higher data rate. This diagram also shows that when lower level codes are used, the upper level codes above them cannot be used.

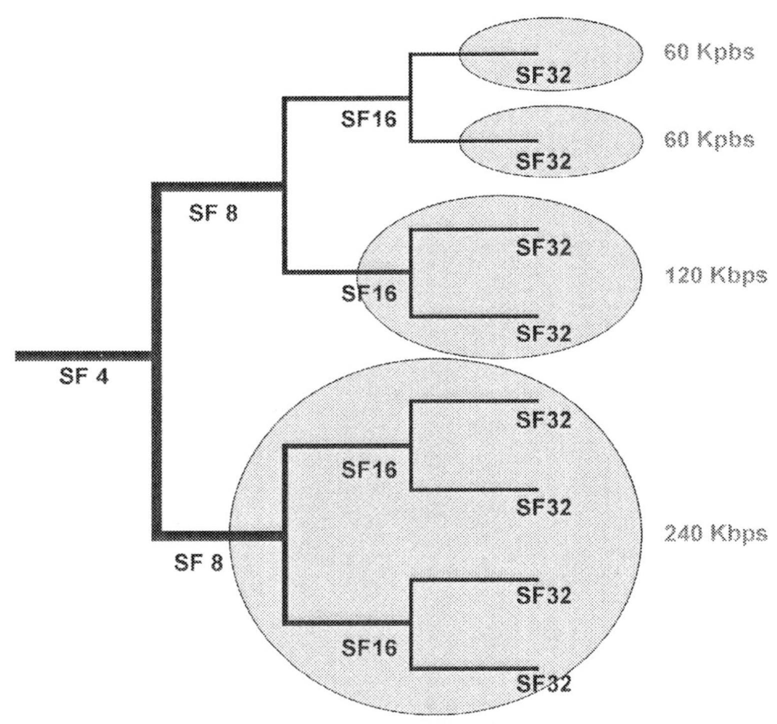

Figure 1.8., Channel Code Tree

Frequency Bands

Frequency allocation is the amount of radio spectrum (frequency bands) that is assigned (allocated) by a regulatory agency for use by specific types of radio services.

It is not possible to use all of the frequency bands assigned to a service provider due to the need for frequency guard bands. Guard bands are a portion frequency that is unused between two channels or bands of frequencies to provide a margin of protection against signal interference. Signal interference may come from radio channels that are operating outside the assigned frequency bands of channels operating near the guard frequency.

In 1992, the frequency bands for UMTS systems were designated at the world administrative radio conference (WARC). The new requirements for the UMTS systems were defined by the international telecommunications union (ITU).

Figure 1.11 shows the frequencies that have been designated for 3rd generation UMTS WCDMA systems. This diagram shows that the designated frequency bands vary throughout the world because some countries have already assigned (licensed) part of the frequency bands.

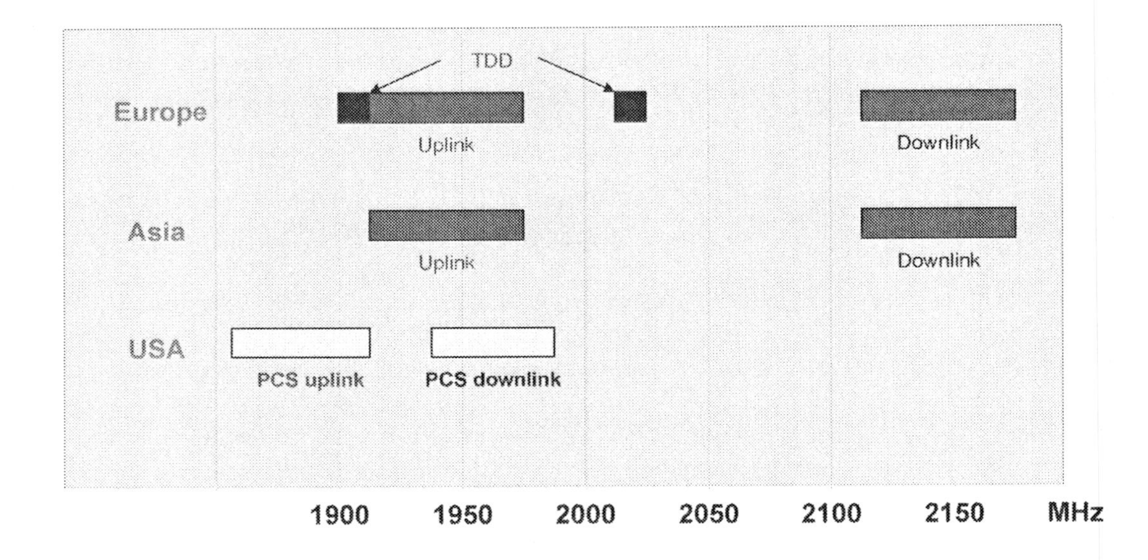

Figure 1.11, WCDMA UMTS Frequency Bands

Frequency Reuse

Frequency reuse is the process of using the same radio frequencies on radio transmitter sites within a geographic area that are separated by sufficient distance to cause minimal interference with each other. Frequency reuse allows for a dramatic increase in the number of customers that can be served (capacity) within a geographic area on a limited amount of radio spectrum (limited number of radio channels). The ability to reuse frequencies depends on various factors that include the ability of channels to operate in with interference signal energy attenuation between the transmitters.

The WCDMA radio channels use coded channels that are uniquely assigned to each user. This allows many users to operate on the same frequency. This also allows frequencies to be reused in every cell site and sectors within a cell site. However, the use of the same frequency in the same cell site and sector increases the interference levels and decreases the capacity of the radio channels.

Figure 1.12 shows how WCDMA systems can reuse the same frequency in each cell site. This example shows that the frequency use factor is 1 (N=1) and that the overlap of the radio channels results in an increased interference level in the overlapping area. Because multiple chips represent each channel, this overlap simply results in the loss of some of the chips and this reduces the capacity of the WCDMA system.

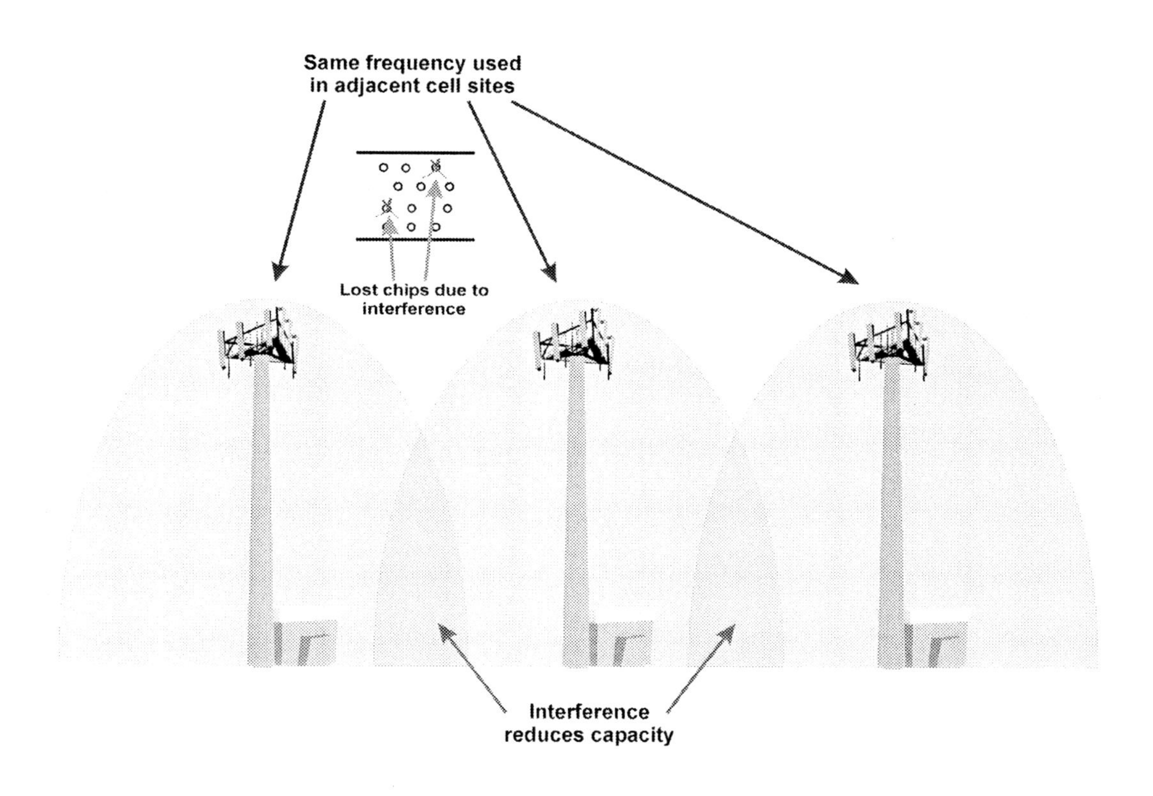

Figure 1.12., WCDMA Frequency Reuse

Frequency Diversity

Frequency diversity is the process of receiving a radio signal or components of a radio signal on multiple channels (different frequencies) or over a wide radio channel (wide frequency band) to reduce the effects of radio signal distortions (such as signal fading) that occur on one frequency component but do not occur (or not as severe) on another frequency component.

Because the WCDMA radio channel provides communication over a relatively wide 5 MHz radio channel (compared to the 200 kHz GSM channel), it is less susceptible to signal fading. When radio signal fades occur (due to signal combining and canceling), they generally occur over a narrow frequency range. This means a signal fade only affects the reception of some of the chips that represent each bit of information that is transmitted. If a majority of the remaining chips can be successfully received, this results in the successful transfer of information, even in the presence of radio signal fades.

Figure 1.13 shows how a wideband radio channel offers the capability of frequency diversity. This example shows that only a portion of the wideband WCDMA radio channel is affected by the radio signal fade. As a result, only a few of the chips are lost and a majority of chips are successfully transmitted.

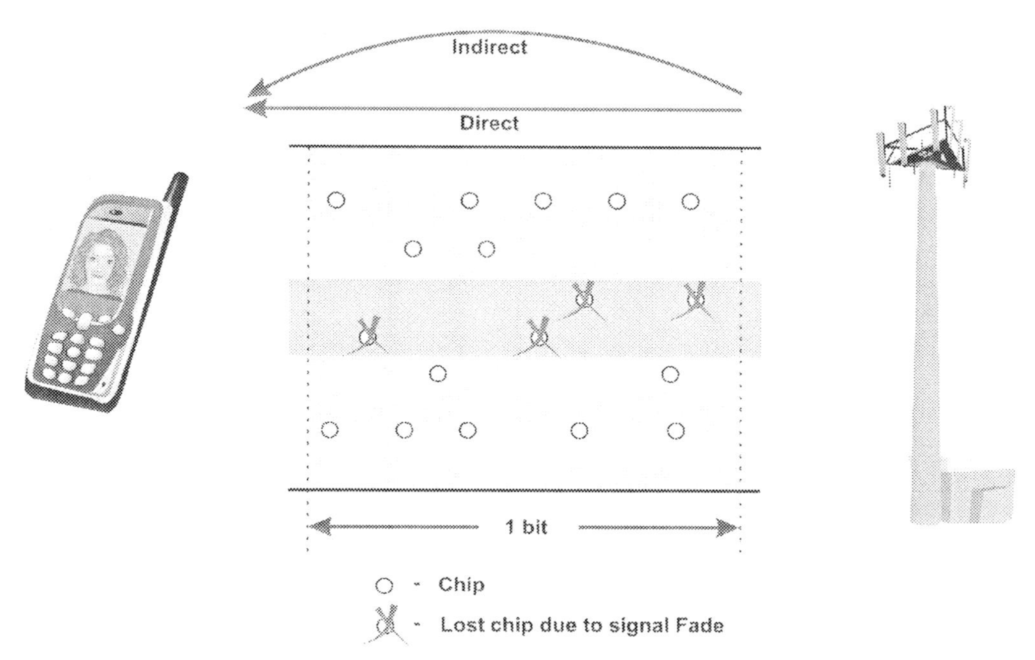

Figure 1.13., WCDMA Frequency Diversity

Time Diversity

Time diversity is the process of sending the same signal or components of a signal through a communication channel where the same signal is received at different times. The reception of two or more of the same signal with time diversity may be used to compare, recover, or add to the overall quality of the received signal.

The cause for time diversity may be naturally created or it may be self-induced. Delayed signals may be created by the reflection of the same signals by objects (such as buildings or mountains). When multiple signals are received that take different paths (direct and/or reflected), it is called multipath. In some cases, it is desirable to self-create multiple delayed signals in the transmitter. The creation of multiple delayed signals can be used to overcome the effects of signal fading. In either case, the WCDMA uses a receiver that is capable of decoding two or more signals that are delayed in time. This receiver is called a rake receiver.

Figure 1.14 shows how a CDMA system can use a rake receiver to combine multiple time delayed (multipath) signals to help produce a higher quality received signal. In this example, the same signal is received at the mobile device because one of the signals has been reflected off a building during transmission. This reflected signal has to travel a longer distance so it is delayed slightly when the mobile device receives it. Because each signal is identified by a unique code, the receiver can separately decode each signal. The receiver can then select or combine the two (or more) reflected signals to produce a higher quality received signal.

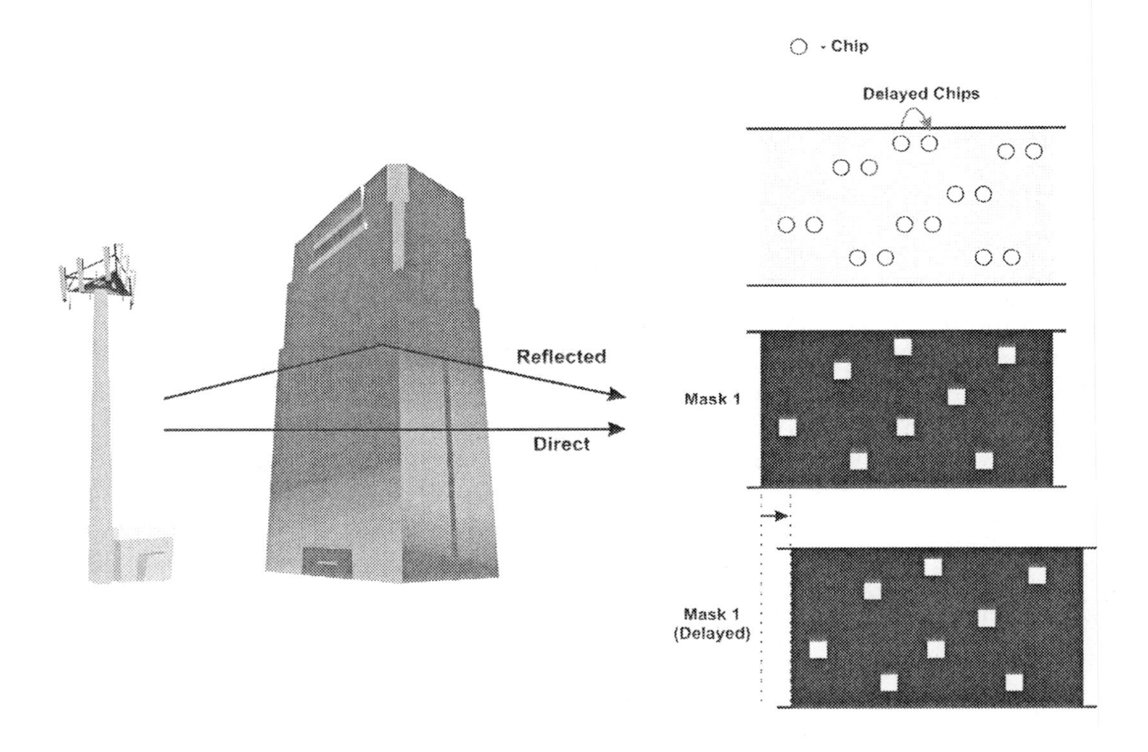

Figure 1.14., WCDMA Rake Reception

Power Control

RF power control is a process of adjusting the power level of a mobile radio as it moves closer to and further away from a transmitter. RF power control is typically accomplished by the sensing the received signal strength level and the relaying of power control messages from a transmitter to the mobile device with commands that are used to increase or decrease the mobile device's output power level.

RF power control is very important to the WCDMA system as the power level that is received from each mobile device must be approximately the same power level as the signals received from other mobile devices. If one mobile device were received at a much higher power level than the other devices, the reception of the lower power channels would be lost in the interference and noise. As a result, more accuracy in power level control can results in received signals of the same level. This results in higher levels of system capacity (more customers can simultaneously operate on the same channel).

To accurately control the RF power level, The WCDMA RF channel uses a combination of course (open loop) power control and precise (closed loop) power control. Open loop power control is a process of controlling the transmission power level for a mobile device using the received power level and an indicator to how much RF power it should transmit. As the received signal level goes down, the transmitter signal level is increased. Open loop power control provides for relatively broad (wide range) power level control.

Closed loop power control is a process of controlling the transmission power level for the mobile device using the power level control commands that are received from another transmitter that is receiving its signal (the closed loop). Closed loop power control in the WCDMA system is used to precisely control (fine tune) the mobile device transmitter level. When in closed loop operation (when the mobile device is assigned to a specific channel), the base station sends power level control commands 1,500 times per second (every 666 usec). The power control commands can change the output power in 1, 2 or 3 dB steps.

Figure 1.15 shows that the WCDMA system uses a combination of open-loop and closed-loop power control to ensure that approximately the same RF signal level is received from all users of the system regardless of the distance they are from the cell site. This example shows how open loop power control automatically adjusts the transmitter power level of the mobile device as the received signal level decreases and how closed loop power control fine tunes to receive signal power level at an antenna by continuously sending small power level control commands.

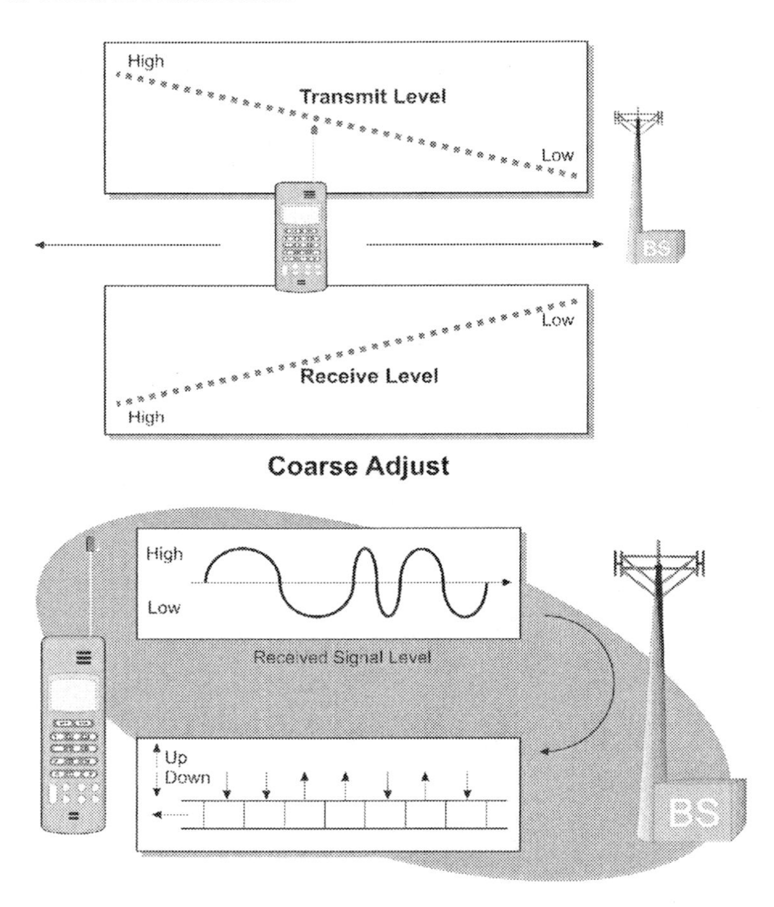

Figure 1.15., Open and Closed Loop WCDMA RF Power Control

RF Power Classification

RF Power classification defines the RF power levels associated with specific modes of operation for a particular class of radio device. Classes of RF devices often vary based on the application and use of the device such as portable, mobile or fixed applications. RF power classification typically defines the maximum RF power level a device may transmit but it may also include the minimum RF power levels and the RF power levels for specific modes of operation (such as during power control commands). The maximum RF power for each class typically determines the type of mobile device; car mounted (high power class), transportable (medium power class), or handheld portable (low power class). WCDMA mobile devices have 4 different power classes 1-4 (most portable mobile telephones are class 4).

Mobile devices that operate close to the base station will receive a high quality signal and this will allow them to transfer at the highest data transmission rates (they can use lower spreading factors). Mobile devices that operate far away from the base station (at the edge of the radio coverage area) will typically be assigned to slower data transmission rates that use higher spreading factors (more protection from interference).

Base stations typically transmit at full power. This allows mobile devices to accurately measure the signal strength of nearby base stations.

Figure 1.16 shows the different types of power classes available for WCDMA mobile devices and how their maximum power level. This table shows that there are 4 classes of WCDMA devices and their maximum power level ranges from +21 dBm (0.125 Watts) to +33 dBM (2.0 Watts).

Mobile Device Class (UE Class)	Maximum Power Level (dB)	Maximum Power Level Watts
1	+33 dBm	2.0 Watts
2	+27 dBm	0.5 Watts
3	+24 dBm	0.25 Watts
4	+21 dBm	0.125 Watts

Figure 1.16., WCDMA User Equipment (UE) Power Classification

Channel Structure

Channels are a portion of a physical communications channel that is used to for a particular communications purpose. There are two groups of channels used in the WCDMA system; control channels and traffic channels. Control channels are used to setup, manage, and terminate communication sessions. Traffic channels are primarily used to transfer user data but can also transfer some control information. Some of the channels on the WCDMA system use separate coding so they can be simultaneously transmitted and decoded by the receiver and other channels are logical channels that share a coded channel.

Channel structure is the division and coordination of a communication channel (information transfer) into logical channels, frames (groups) of data, and fields within the frames that hold specific types of information. The WCDMA system has different channel structures for the forward and reverse directions.

Figure 1.17 shows the channel structure and duplex channel spacing for the WCDMA system. This diagram shows that each coded communication channel (traffic channel) is divided into 10 msec frames an that each of the frames is divided into 666 usec time slots. All the time slots are used during normal transmission. This example also shows that the duplex channel spacing between the uplink and the downlink is nominally 190 MHz (85 MHz for USA). The 10 msec frames for the forward and reverse channels is transmitted with a fixed time offset relative to each other.

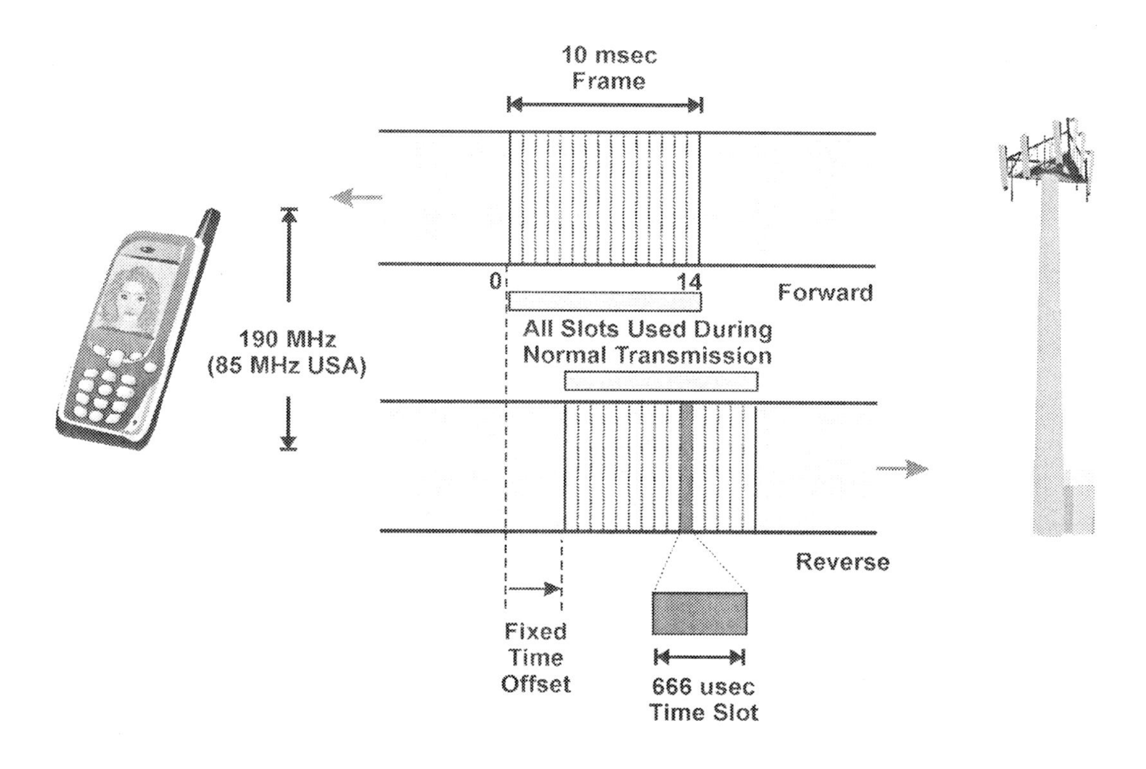

Figure 1.17., WCDMA Duplex Channels

Packet Data Transmission

Packet data transmission is the process of setting up and managing packet data communication sessions. There are three basic ways to send packet data on the WCDMA system; on the access channel, on a common paging channel, or on a dedicated packet data channel.

If the amount of information (packet data) is very small, it can be sent directly on the random access channel (RACH). The sending of very short data packets on the access channel eliminates the need to setup another communication channel for more packets.

If the amount of information is larger than can be sent on the RACH channel, the data can be sent on the common paging channel (CPCH) channel (shared by other users). If the amount of data is relatively large, the dedicated data channel (DCH) is used (not shared by others).

Packet data transmission starts by sending access request messages at low RF power level and then gradually increasing the RF power level of the access request messages until the system responds to the access request or when a maximum allowable RF power level has been reached. If the WCDMA system does respond to the access requested within a short period of time, the mobile device must stop transmitting and wait a random amount of time before attempting to send access request messages to the system again.

Access request messages contain a preamble that alerts the system that an access request message is coming along with a short identification number that is used to identify the mobile device that is attempting to access the system. If the system recognizes the access request message, it will send an acknowledgement message on the CPCP AP-AICH channel.

When the mobile device receives the acknowledgement message, it can continue to transmit additional access control information that is necessary to setup the packet data transmission session. These control messages are confirmed on the CD/CA-ICH channel. The CD/CA-ICH channel is used to coordinate the shared packet data channel.

If the amount of data to be sent is very large or the data transmission is likely to be continuous (e.g. multimedia streaming), a dedicated data channel (DCH) can be setup. The sending of data on dedicated channels does involve additional setup time. However, the transmission of packet data on DCH allows the system to provide high-speed data transmission and to allow for uninterrupted soft handovers for data sessions.

Packet data transmission is controlled by a packet scheduler (PS). The PS coordinates the distribution of packets over multiple radio channels, frames, and time slots. The packet scheduler is commonly located at the RNC to allow a PS to coordinated packet data transmission over several cells.

Figure 1.18 shows how the WCDMA system transfers packets on the CPCH. This diagram shows that the mobile device first attempts access to the system on the CPCH. If this request is acknowledged on the AP-AICH, the mobile device continues to transmit packet data access request information. If the mobile device does not hear a response to its access request message, it will increase its transmitter power level and send its access probe again. This process repeats until either the WCDMA system acknowledges its request or the mobile device reaches its maximum allowable transmit power. This example shows that the WCDMA system can control the power level on the CPCH channel.

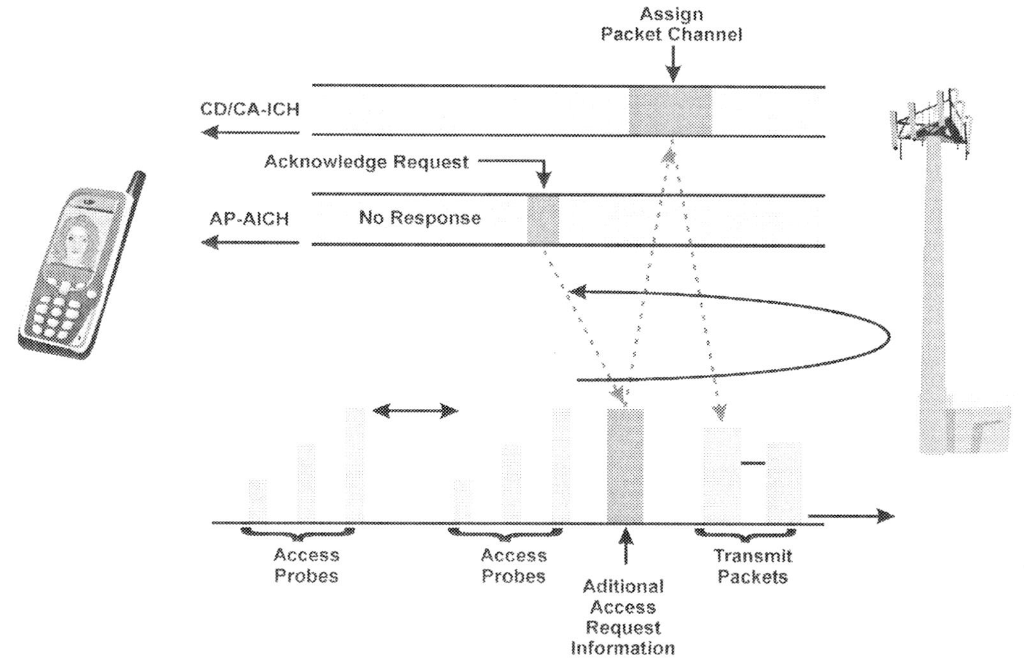

Figure 1.18., WCDMA Packet Access on the CPCH Channel

Asymetric Channels

Asymetric channels are two-way communication channels that allow for transmission rates that can vary by direction. For example, the downlink broadcast channel may be a high-speed channel (e.g. 1.9 Mbps) and an uplink (reverse direction) channel may only be 15 kbps. The WCDMA system allows asymmetric operation as it permits the assignment of different data transmission rates for the forward and reverse directions.

Speech Coding

Digital speech compression (speech coding) is a process of analyzing and compressing a digitized audio signal, transmitting that compressed digital signal to another point, and decoding the compressed signal to recreate the original (or approximate of the original) signal.

Figure 1.19 shows the basic digital speech compression process. The first step is to periodically sample the analog voice signal (20 msec) into pulse code modulated (PCM) digital form (usually 64 kbps). This digital signal is analyzed and characterized (e.g. volume, pitch) using a speech coder. The speech compression analysis usually removes redundancy in the digital signal (such as silence periods) and attempts to ignore patterns that are not characteristic of the human voice. In this example, this speech compression process uses pre-stored code book tables that allow the speech coder to transmit abbreviated codes that represent larger (probable) digital speech patterns. The result is a digital signal that represents the voice content, not a waveform. The end result is a compressed digital audio signal that is 8-13 kbps instead of the 64 kbps PCM digitized voice.

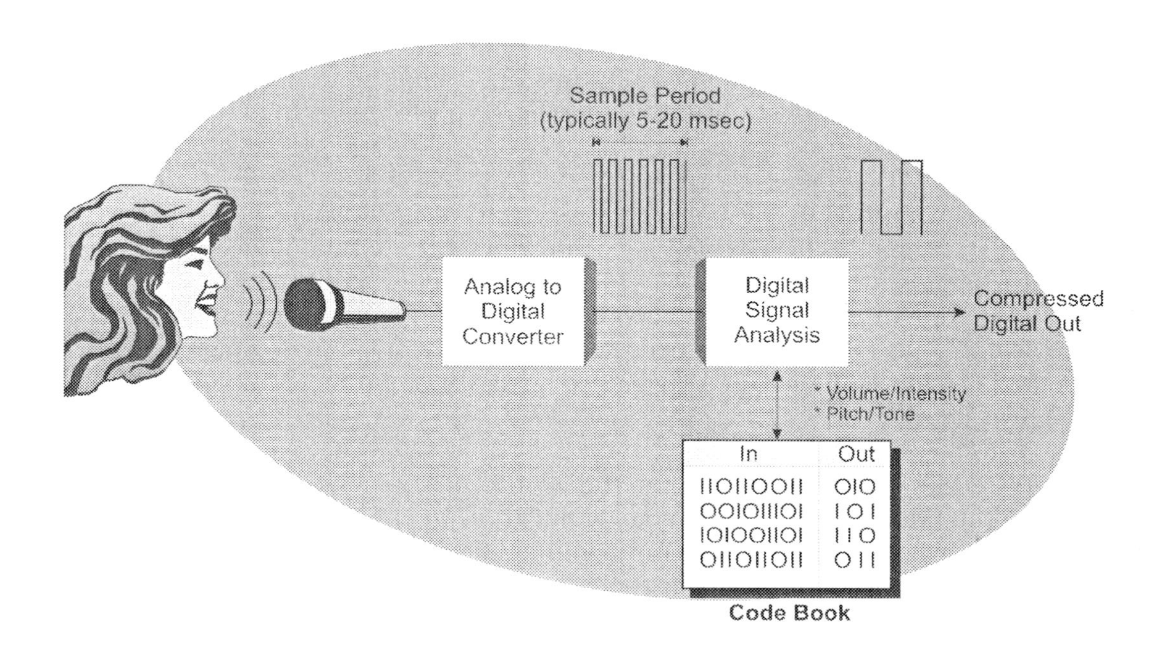

Figure 1.19., Speech Coding

The WCDMA system is designed to use different types of speech coders depending on the system capability and the needs of the user. The ability of the WCDMA system to change speech coder technologies is called adaptive multi-rate (AMR) speech coding.

The use of speech coders that have different compression rates allows a lower bit rate coding process (higher compression rates) to be used when system capacity is limited and more users need to be added to the system.

The WCDMA system speech coders include the 12.2 kbps standard GSM enhanced full rate (EFR), 7.4 kbps interim standard 641 (IS-641), 6.7 kbps Pacific digital cellular (PDC) EFR, and other types of speech coders that operate at 10.2 kbps, 7.95 kbps, 5.9 kbps, 5.15 kbps, and 4.75 kbps. In the future, the WCDMA system may add other types of speech coders as technology improves.

The determination of the speech coder that is used is a combination of mobile device capability (which speech coders the mobile device has available to use), system capability (which speech coders the base station has available to use), and which speech coder is preferred by the system (lower bit rate speech coders may be used when systems reach overcapacity conditions.) The selection of which speech coder is used can be changed every 20 msec.

The WCDMA system provides for the ability of the system to dynamically change the maximum data transmission rate that the speech coder may use to characterize (compress) the audio signal. While this maximum data rate limit may affect the quality of speech (less bits may mean lower audio quality in periods of high audio activity), the limit allows the WCDMA system to control the maximum data transmission rate and may allow the system to add additional users with a tradeoff of audio quality (called soft capacity).

Figure 1.20 shows the basic adaptive multi-rate speech coding process. This diagram shows that the WCDMA system allows for the use of different speech coding processes and that the system may dynamically control which speech process is used dependent on the needs of the system. This example shows that the speech coding process can change as often as every 20 msec speech frame. The selection of speech coder is primarily determined by the channel quality. However, this example shows that the system can instruct the speech coder to use high compression process (lower audio quality) during periods of system overcapacity.

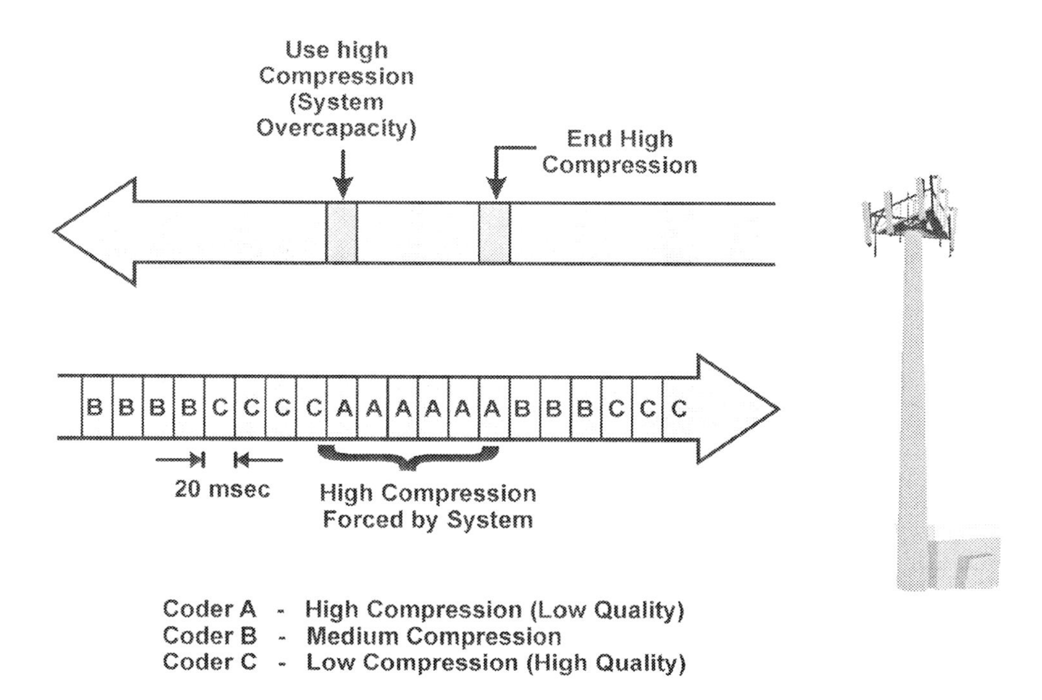

Figure 1.20., WCDMA Adaptive Multi-Rate Speech Coding

The speech coding process can use a process of error concealment to reduce the effect of errors that are received during transmission. Error concealment is a process that is used by a coding device (such as a speech coder) to create information that replaces data that has been received in error. Error concealment is possible when portions of the signal output of the coder has some relationship to other portions of the signal output and that the relationship can be used to produce an approximated signal that replaces the lost information period (lost bits).

To save battery life, the WCDMA system can use voice activity detection (VAD) to reduce or stop transmission during periods of inactive speech activity. The inhibiting of transmission is called discontinuous transmission (DTx). Voice activity detection (VAD) is a process or an electronic circuit that senses the activity (or absence) of voice communication signals. When the VAD system determines the user has stopped talking, the transmitter may be shut off. The use of VAD and DTx reduces the power consumption (e.g. extends the battery life) and reduces interference to other users (increasing system capacity). To overcome the potential disconnection from the system, a silence descriptor (SID) frame may be sent to allow the system to know the connection is still active and to provide some background noise so the listener knows the call is still in progress (not complete silence).

Soft Capacity Limits

A capacity limit is the maximum amount of service (such as data transmission rate) or number of customers that a system can provide services to at a defined level of service. The WCDMA system has a soft capacity limit as the system operator can dynamically change the defined level of service to change the maximum number of customers (capacity limit) that can obtain service from the system. This allows the service provider to temporarily increase the system capacity in exchange for a reduction in the quality of voice.

A WCDMA service provider can increase the number of customers on a WCDMA mobile communication system by reducing the audio quality through increased speech compression. This lowers the average data rate per user, reducing interference, increasing the maximum number of users.

Figure 1.21 shows that a soft capacity limit allows for the gradual decay of voice quality in a communication system when additional users are added in a system. To provide service to more customers than the capacity limit (over capacity) in a WCDMA system, users in the system are provided with lower bit rates (higher speech compression). As a result, assigning lower bit rates to users as service demand increases trades off voice quality for increases in system capacity.

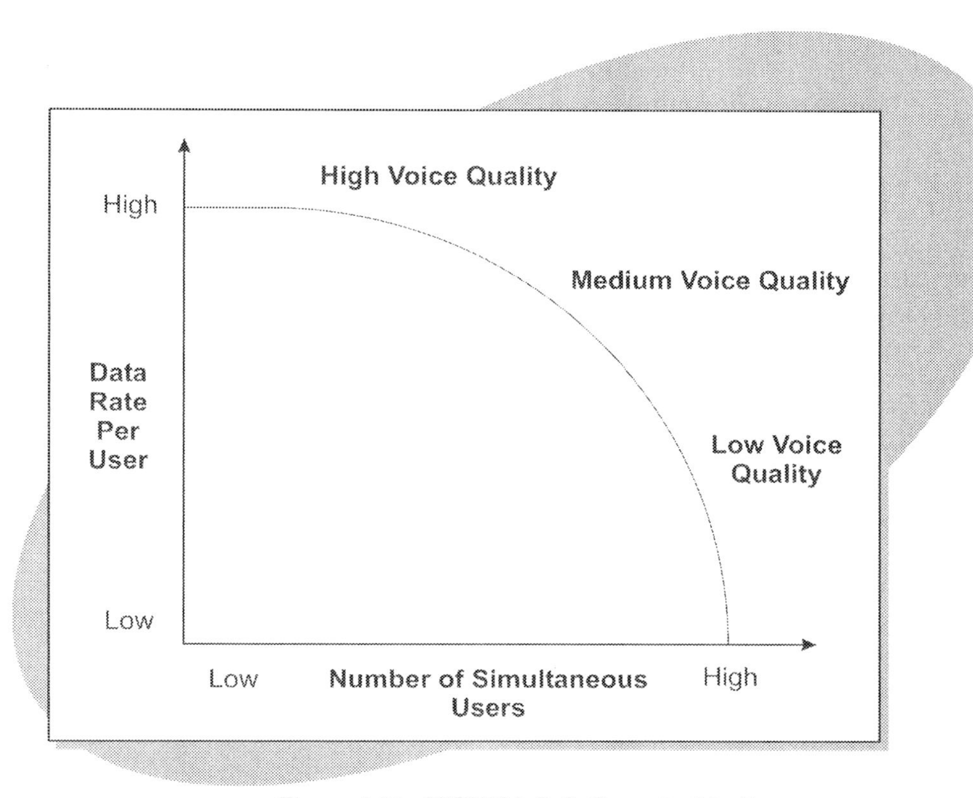

Figure 1.21., WCDMA Soft Capacity Limit

Handover

Handover (also called handoff) is a process where a mobile device that is operating on a particular channel is reassigned to a new channel. This can be a new frequency channel, new code channel, or a new logical channel. The process is often used to allow subscribers to travel throughout the large radio system coverage area by switching the calls (handoff) from cell-to-cell (and different channels) with better coverage for that particular area when poor quality conversation is detected.

Handoff may also occur when a mobile device requests a service that can only be provided by a radio channel that has different service capabilities. This might mean assignment from a WCDMA traffic (voice) channel to a GSM traffic channel.

The WCDMA system can handover a communication channel when the radio network controller (RNC) determines the radio channel quality between the mobile device and the base station has fallen below an acceptable level and a better radio channel is available for call transfer. The RNC can use information that it receives from the mobile device to assist in its' handover decisions. The mobile device can measure the received signal strength (RSSI) and the received signal code power (RSCP) level and return this information to the RNC via the base station. The WCDMA system handover types include soft handover, inter-system handover, and hard handover.

Soft Handover

A soft handover is a process that maintains a communication connection with an initial transmitter site (e.g. base station) while simultaneously communicating with one or more additional transmitter sites (base stations) during the handover process.

The WCDMA system uses soft handover process to allow the mobile device to be precisely controlled during the handover (cell site transfer) process and to provide for undetectable changes in audio quality during handover. During the soft handover process, two or more cell sites are simultaneously communicating with the mobile device during the transfer process. The information received from these cell sites

To begin the soft handover process, the mobile device monitors the signal strength of pilot channels of nearby cell sites and returns this information to the cell site it is communicating with. This list of pilot channels is provided on the cell site's broadcast channel.

Figure 1.22 shows the basic soft handover process that occurs in the WCDMA system. In this example, the mobile device has measured the signal strength of the pilot channel from a nearby cell site. Using this information, it is determined that the adjacent cell site is a candidate for the soft handover process and that it should simultaneously communicate with the mobile device and the soft handover process begins. During the soft handover process, both cell sites simultaneously receive information from the mobile device and the information frames that have the highest quality (least errors) can be selected.

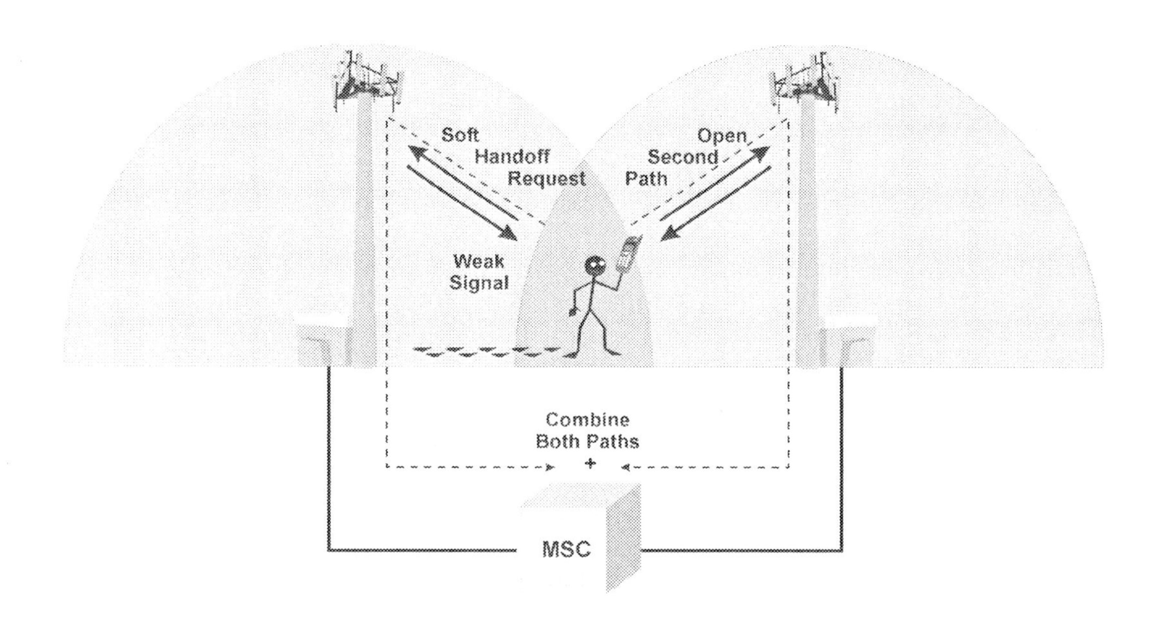

Figure 1.22., WCDMA Soft Handover

Inter-System WCDMA to GSM Handover

Mobile devices that are communicating on a WCDMA channel may need to handover to a GSM cellular channel. Some of the reasons this transfer may be necessary include no availability of WCDMA radio channels during handover or the need to use GSM services that are not available on the WCDMA system.

This may occur because the geographic area the mobile device is entering into does not have WCDMA radio channels. Some system operators may only deploy (install) WCDMA channels or service features (such as packet data) on part of the wireless system.

Transferring calls between a WCDMA system and GSM system has some unique challenges that include different frequencies and time slot periods. If the WCDMA mobile device is continuously transmitting a data signal, this would require the mobile device to simultaneously communicate with both the GSM system and the WCDMA system (2 receivers). To overcome this potential challenge, the WCDMA system can use a compressed mode (formerly called "slotted mode") of transmission. This allows the mobile device to adapt its data transmission rate for brief periods to allow it to tune (change its frequency) to monitor and communicate if necessary with a GSM channel in between WCDMA transmissions. The compressed mode can be entered into by either lowering the user data rate, changing the spreading factor (increasing the WCDMA channel data transmission rate), or by puncturing the error protection coding (using less error protection bits).

Hard Handover

Hard handovers are the transferring of communication session from one communication channel (frequency and/or coded channel) to a new communication channel where one channel is completely disconnected before a new channel is established. During the hard handover process, small amounts of information or audio are lost during the transition process.

WCDMA hard handovers occur when a call is transferred (handed over) from a WCDMA system to a WCDMA channel that is operating on another frequency or when it is handed over to another WCDMA channel that is operating on another WCDMA system.

When the handoff is occurring between different systems, the mobile telephone can enter into a compressed transmission mode. During the compressed transmission mode, speech and/or data information are sent during a portion of the frame and other portions of the frame are used to allow the mobile device to search, find, and measure other candidate radio channels for handover.

Radio Coverage and Capacity Tradeoff

Radio coverage is a geographic area that receives a radio signal above a specified minimum level. The radio coverage area of a WCDMA cell site can change based on the capacity used by customers. As more customers are added to a radio coverage area, the smaller the radio coverage area becomes (given the same power levels).

Figure 1.23 shows how the WCDMA system can tradeoff radio coverage for higher capacity. This diagram shows as more customers are added to the cell, the radio boundaries begin to contract into a smaller radio coverage area.

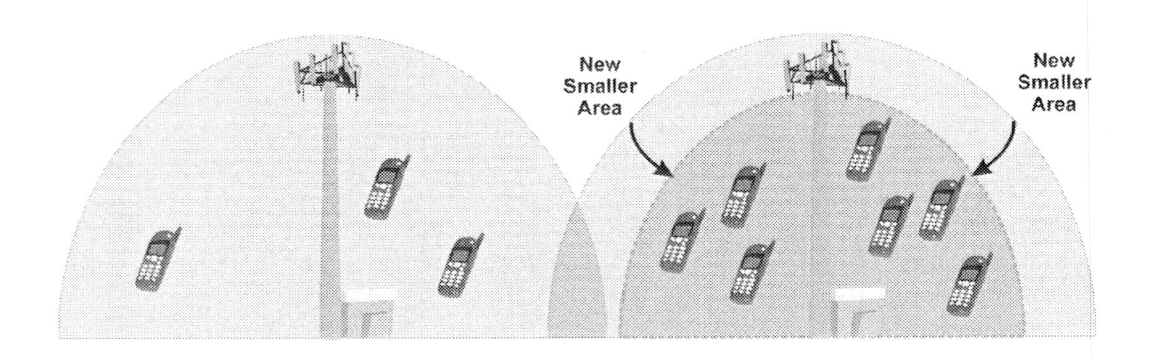

Figure 1.23., WCDMA Radio Coverage and Capacity Tradeoff

Discontinuous Reception (DRx)

Discontinuous reception (DRx) is a process of turning off a radio receiver when it does not expect to receive incoming messages. For DRx to operate, the system must coordinate with the mobile radio for the grouping of messages. The mobile device will wake up during scheduled periods to look for its messages. This reduces the power consumption that extends battery life. This is sometimes called sleep mode.

The WCDMA system divides the paging channel into sub-channel groups to provide for DRx capability. The number of sub-channel groups is determined by the system. Each 10 frames contain a paging channel frame. To inform the mobile device of the sleep periods, a paging indicator channel (PICH) is used. A paging indicator (PI) message is sent at the beginning of the paging channel frame to identify the paging channel group. This allows the mobile device to quickly determine if it must keep its receiver on during the paging group or if it can turn off its receiver and wait for the next paging channel group.

Figure 1.24 shows the basic DRX (sleep mode) process used in the WCDMA system. This diagram shows that each 10 msec frame contains a paging channel frame and that the beginning of the paging channel frame contains a paging indicator (PI) message. The PI message identifies the paging group and this allows the mobile device to determine if it needs to keep it's receive on (wake mode) or if it can turn off its receiver (sleep mode) until the next group of paging messages.

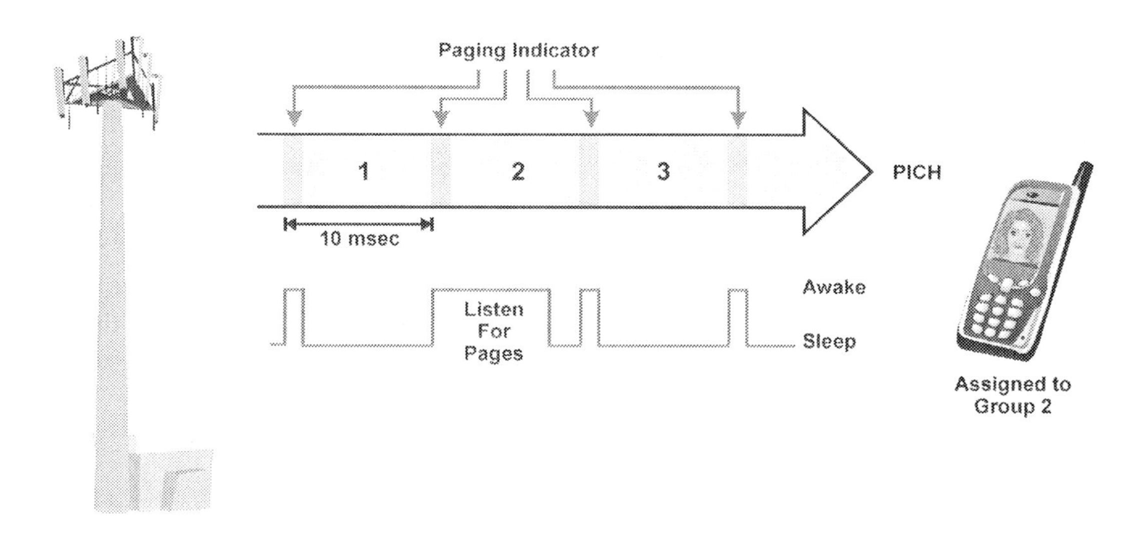

Figure 1.24., Discontinuous Reception (Sleep Mode)

WCDMA Radio Channels

A radio channel is a communications channel that uses radio waves to transfer information from a source to a destination. A radio channel may transport one or many communication channels and communication circuits.

The WCDMA system uses the same 5 MHz bandwidth size radio channels for WCDMA voice and WCDMA data. To maximize the data transfer rate, the WCDMA forward radio channel can use different types of modulation and different amounts of signal spreading (processing gain). The forward WCDMA channel transmits at maximum power to one user at a time while the reverse WCDMA channel uses precise power control and allows all the active (connected) users to transmit at the same time.

To create a wide digital radio channel from low-speed digital audio or control signals, each bit of the information signal is multiplied (converted) into a long sequence of bits called a spreading code. This spreading code effectively represents each bit of an information signal by multiple RF information signals (chips) over a frequency band that is much wider than the information signal. This is why it is called a wideband system.

Because the WCDMA system use variable spreading sequence for each coded communication channel, data transmission rates can vary based on the amount of spreading that is used. For the user data channels, the information data transmission rate can vary from 7.5 kbps to 960 kbps. While it is theoretically possible to obtain these data transmission rates, the total data transmission rate also varies (is reduced) based on the amount of interference received from other channels.

The WCDMA system can uses a variety of speech compression (speech coding) devices that have different data transmission rates. The ability to select speech coders with a higher data compression capability allows the service providers to increase the number of customers per channel in exchange for the reduction of voice quality (increased speech compression typically decreases the voice quality). This capacity to quality tradeoff is known as the soft capacity limit.

Figure 1.25 shows how the WCDMA digital radio channel spreads each information bit into multiple radio information chips. This diagram shows that each information bit is converted (multiplied) by a spreading code. Receivers look for a specific spreading code until a match occurs. If a majority of the chips match, the receiver can recreate the original bit of information.

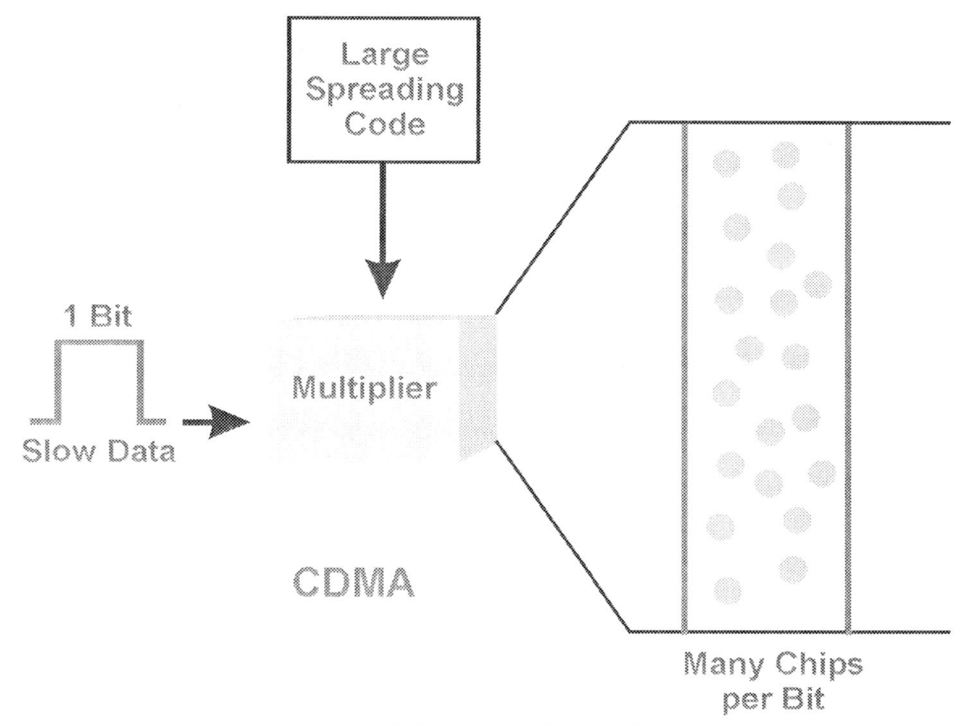

Figure 1.25., WCDMA RF Channel Spreading

The physical channel is divided into 10 msec frames and each frame is divided into 15 time slot periods (666 usec each). During typical communication with the system, a mobile device is assigned to one or more time slots during a frame (multiple time slots are combined to achieve higher data transmission rates). Some of the time slots are used to transfer control information and some of the slots are used for user information (voice and/or data).

Channel Bandwidth

Radio channel bandwidth is the difference between the upper frequency limit and lower frequency limit of allowable radio transmission energy for a radio communication channel. The WCDMA system uses 5 MHz radio channels for both voice and data communication. For the FDD WCDMA system, two radio channels are combined (paired) to form a full duplex (simultaneous two-way) communication channel. For the TDD WCDMA system, a single radio channel is used and different time slots in each frame are used to each direction to provide two-way communication.

Modulation Types

Modulation is the process of changing the amplitude, frequency, or phase of a radio frequency carrier signal (a carrier) to change with the information signal (such as voice or data). Adaptive modulation is the process of dynamically adjusting the modulation type of a communication channel based on specific criteria (e.g. interference or data transmission rate).

The WCDMA system uses different modulation types for the downlink and uplink channels. The downlink channel uses phase shift keying (QPSK) and the Uplink uses dual channel modulation, one channel is used for the control channel and the other channel is used for the user traffic (data) channel.

Because all the downlink communication channels are controlled by a base station and the radio channel transmission is continuous. The data transmission on the uplink may be cycled on and off during discontinuous transmission (DTx) to save battery energy.

Channel Types

The WCDMA system has 3 basic channel types; physical channels, transport channels, and logical channels. Channels can be classified as common channels or dedicated channels. Common channels are accessible and shareable by a variety of communication devices. The WCDMA system uses common channels to send commands or instructions such as system or device identification information to all mobile devices operating in a radio coverage area. Dedicated channels are accessible by one or several designated devices. The WCDMA system uses dedicated channels to send information to a specific device (such as voice or data information).

Physical Channels

Physical channels are the electrical, radio, or optical transmission channels that are connected between transmitters and receivers. Physical channels may be distinguished from other channels by the frequency, code, or time of occurrence of the transmission. The WCDMA system is composed of several physical (coded) and logical (time shared) channels. WCDMA channels may be only used on a forward (downlink) or reverse (downlink) direction or they may be paired for two-way transmission.

Common Pilot Channel (CPICH)

The common pilot channel (CPICH) provides a reference timing signal to assist in the demodulation of the received signal. The CPICH uses a predetermined spreading code that is unique from all other channels. Because the CPICH is always present, it can also be used by mobile devices to estimate the received signal strength. The CPICH has a fixed spreading factor of 256 that provides a data transmission rate of 30 kbps.

The WCDMA system uses two types of pilot channels; a primary channel and a secondary channel. The primary channel code is used for each cell (or sector). A secondary pilot channel that uses other codes may be included to assist with the use of directional (smart) antenna systems.

Primary Common Control Physical Channel (PCCPCH)

The primary common control physical channel (PCCPCH) is used to continuously broadcast WCDMA system identification and access control information on the forward (downlink) channel. The PCCPCH uses a fixed spreading code of 256 and uses ½ rate convolution coding for error protection that provides a data transmission rate of 30 kbps. PCCPCH information is also interleaved (distributed) over two consecutive 10 msec frame periods.

Physical Downlink Shared Channel (PDSCH)

The physical downlink shared channel (PDSCH) is used to send common control information to all mobile devices operating within its coverage area. The PDSCH can use a spreading factor that ranges from 4 to 256 and the PDCH is always associated with a downlink dedicated channel (DCH).

Secondary Common Control Physical Channel (SCCPCH)

The secondary common control physical channel (SCCPCH) provides common access control information and call alert (paging) messages to mobile devices that are operating within its area. The spreading factor for the SCCPCH is determined by the capabilities of the mobile device reception capability. The SCCPCH is composed of the forward access channel (FACH) and a paging channel (PACH).

A WCDMA system must have at least one SCCPCH. Additional SCCPCHs can be added to increase the capacity of the system to process system access requests and to send more paging messages.

Physical Random Access Channel (PRACH)

The physical random access channel (PRACH) is an uplink channel that is used to coordinate the random requests for service from mobile devices. The PRACH transmits access requests (bursts) when a mobile device desires to access the mobile system (call origination or a paging response). To assist the base station in receiving and decoding the access request messages, the beginning part of the PRACH access request message uses a fixed spreading factor of 256 and the data transmission rate of the PRACH is 16 kbps. The message part of the PRACH message can use a spreading factor from 32 to 256 kbps.

Dedicated Physical Data Channel (DPDCH)

The dedicated physical data channel (DPDCH) is used to transfers user data between the mobile device and the base station (uplink and downlink channels). The DPDCH can use different spreading factors for the uplink and downlink DPDCHs . The downlink DPDCH can use a spreading factor that ranges from 4 to 256 and the uplink DPDCH can use a spreading factor that ranges from 4 to 512. The spreading factor can change on each frame.

The uplink DPDCH uses dual channel I/Q (phase) modulation to send information from two different data sources on each I/Q channel. One phase code is used to send user data and the phase code is used for sending control signals. Using dual channel modulation allows the DPDCH to simultaneous send data and control information.

Dedicated Physical Control Channel (DPCCH)

The dedicated physical control channel (DPCCH) is used to send control information between the WCDMA system and specific mobile devices. Both the uplink and downlink DPCCH carries pilot bits and transport format combination identifier (TFCI) and the downlink DPCCH also include feedback information bits (FBI) and transmission power control (TPC) bits. The pilot bits are used as a reference signal to help demodulate the signal and to estimate the received signal strength of the channel. The TFCI indicates if multiple physical channels have been combined (to achieve higher data transmission rates). The TPC is used to control the power of the mobile device and the FBI bits are used to assist with diversity transmission.

Physical Common Packet Channel (PCPCH)

The physical common packet channel (PCPCH) is used to transfer packet data and its operation is similar to the random access channel (RACH). When a mobile device wants to send packet data, it must first gain the attention of the system and compete for access on the PCPCH. Like the RACH, the mobile device first monitors the system to see if it is not busy and then it transmits brief access request bursts, gradually increasing the power of each access burst until the system recognizes and responds to its request. After the mobile device has gained the attention on the system, the PCPCH is used to continue the transmit data on the common channel.

Synchronization Channel (SCH)

The synchronization channel (SCH) provides information that assists the mobile device to find and time-align with the system. The WCDMA has primary and secondary synchronization channels. The primary synchronization channel uses a fixed 256 chip sequence that is used in every cell. The secondary synchronization channel provides information that allows for frame and time slot timing synchronization for each base station.

Acquisition Indication Channel (AICH)

The acquisition indication channel (AICH) assigns a mobile device to a data channel (DCH). The AICH is used to send channel assignment messages after the mobile device has competed for access to the system on a random access channel (RACH).

Page Indicator Channel (PICH)

The page indicator channel (PICH) coordinates the sleep mode of the mobile device's receiver section. The PICH works on combination with the paging channel (PCH) to inform the mobile device when it can "sleep" and when it must "wake-up" to receive its paging messages.

Transport Channels

Transport channels are communication channels that use one or more physical channels in a specific way (such as specific channel codes) to transfer information. Transport channels define how, when, and which physical channels are used.

Random Access Channel (RACH)

The random access channel (RACH) is used to initiate requests for service from the mobile device to the base station (uplink). Responses to service requests received on the RACH are provided on the acquisition indicator channel (AICH).

A key feature of the RACH is its ability to transfer small amounts of data with along the system access request message. For applications that send very small amounts of data (such as an electric meter reading), this is a very efficient use of packet data transmission as the sending of the data with the access request eliminates the need for an assignment to a dedicated packet data transmission channel.

Common Packet Channel (CPCH)

Common packet channel (CPCH) is used to send packet data from the mobile device to the base station (uplink channel). The CPCH is used for packets that are too large to send directly on the RACH. The CPCH includes RF power level control (the RACH does not include RF power level control) making it suitable to send a large number of packets with minimal interference to other users.

Forward Access Channel (FACH)

The forward access channel (FACH) provides control and data messages to mobile devices that have registered with the system. There must be at least one FACH in each cell (there can be more than one FACH per base station) and it must have a low data transmission rate to allow all mobile devices to receive control and data messages (some mobiles may be capable of higher data transmission rates).

Downlink Shared Channel (DSCH)

The downlink shared channel (DSCH) is a downlink channel that uses a packet scheduling system to dynamically assign time slots for specific users who have packets to transmit or receive. The packet scheduling system selects the frequency (RF channel), chooses a channel code, and assigns time slot(s).

Uplink Shared Channel (USCH)

The uplink shared channel an uplink channel on the WCDMA system allows several user devices to send control or user data on a shared channel. The USCH is only used in the TDD mode.

Broadcast Channel (BCH)

The broadcast channel (BCH) continuously transmits system and access control information to mobile devices that are operating within its radio coverage areas. This information typically includes the system identification code and name, lists of neighboring radio channels, system configuration information, available random access channels and the scrambling codes they use.

Each base station must have a broadcast channel. The broadcast channel continually transmits and this allows mobile devices to examine and measure the signal strength of each broadcast channel at nearby cell sites. The ability of mobile devices to measure broadcast channels allows it to determine which cell site is best suited for communication.

Paging Channel (PCH)

The paging channel (PCH) is used to send messages (page messages) that alert mobile devices of an incoming telephone call (voice call), request for a communicate session (data session), or to request a maintenance service (e.g. location registration update).

In addition to sending paging messages, the paging channel is also used to provide information about discontinuous reception (DRx) that allows the mobile device to turn off its circuitry (sleep) during periods between paging groups.

High Speed Downlink Shared Channel (HS-DSCH)

The high speed downlink shared channel allows multiple devices to share a high speed communication channel through the assignment of specific codes from a common pool of codes that are available for the channel.

Dedicated Channel (DCH)

A dedicated channel is an uplink or downlink communication channel that is only accessible by one device.

Logical Channels

Logical channels are a portion of a physical communications channel, which is used for a particular (logical) communications purpose. The WCDMA system defines logical channels that are related to (mapped onto) the physical channels. Some logical channels may be dynamically assigned (mapped) to more than one physical channel depending on the capability and needs of the system.

Broadcast Control Channel (BCCH)

The broadcast control channels constantly transfer parameters needed by mobile devices to identify and gain access to the communication system.

Paging Control Channel (PCCH)

The paging control channel is used to send page messages to mobile devices to alert them of an incoming telephone call (voice call) or a request for a communicate session (data session).

Common Control Channel (CCCH)

The common control channel is a logical channel that is used for the establishment and maintenance of communication links between mobile devices and node B radio equipment.

Dedicated Control Channel (DCCH)

A dedicated control channel is used to coordinate and control specific mobile devices.

Shared Common Control Channel (SHCCH)

The shared common control channel is a logical channel that is used for the establishment and maintenance of communication links between mobile devices and node B radio equipment. SHCCH are only used in TDD mode.

Dedicated Traffic Channel (DTCH)

A dedicated traffic channel is an uplink or downlink communication channel that is only accessible by one device to transfer user data.

Common Traffic Channel (CTCH)

A common traffic channel is an uplink or downlink communication channel that is shared by several mobile devices for the transfer of user data.

Figure 1.26 shows the mapping (assignment) of logical and transport channels onto physical channels. This diagram shows that some logical channels can be assigned to one or more physical channels dependent on the channel and system needs.

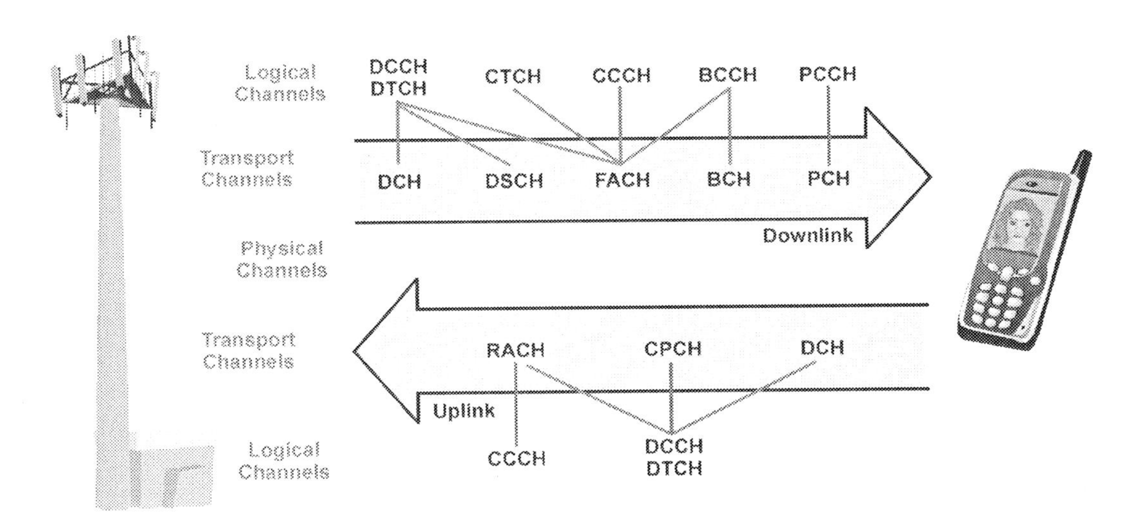

Figure 1.26., Mapping of Transport (Logical) Channels to Physical Channels

WCDMA Network

The WCDMA system includes a Base Stations (BS), various data processing functions, and a data distribution network. The WCDMA network is also called a radio access network (RAN).

WCDMA network parts may be dedicated equipment assemblies or software programs that run on computer servers. WCDMA networks inter-connect wireless devices with nearby radio towers (base stations – called node B) that route calls through switching systems to other wireless telephones to other telephones or data networks. Creating and managing a wireless network involves equipment selection and installation, implementation methods, inter-connection to the public switched telephone network (PTSN) and other networks such as the Internet.

The WCDMA system uses either a mobile switching center (MSC) for voice and medium-speed data or a packet data service node (PDSN) for packet data services. The MSC coordinates the overall allocation and routing of calls throughout the wireless system. The PDSN ensures packets from the mobile devices can reach their destination.

Figure 1.27 shows a simplified functional diagram of a WCDMA network. This diagram shows that the WCDMA system is composed of 3 key parts; the user equipment (UE), UMTS terrestrial radio access network (UTRAN), and a core interconnecting network (CN). The UE is divided into 2 parts, the mobile equipment (ME) and the UMTS subscriber identity module (USIM) card. The UTRAN is composed of base stations (called Node B) and radio network controllers (RNCs). This example shows that the RNCs connect voice calls to mobile switching centers (MSCs) and connect data sessions to packet data service nodes (PDSNs). The core network is basically divided into circuit switched (primarily voice) and packet switched (primarily data) parts. The core network circuit switch parts contain the serving MSC and a gateway MSC. The serving MSC (SMSC) connects to the UTRAN system and the gateway MSC (GMSC) connects to the public telephone network. The core network packet switched parts contain the serving general packet radio service (GPRS) support node (SGSN) and a gateway GPRS service node (GGSN). The SGSN connects to the UTRAN system and the GGSN connects to data networks such as the Internet.

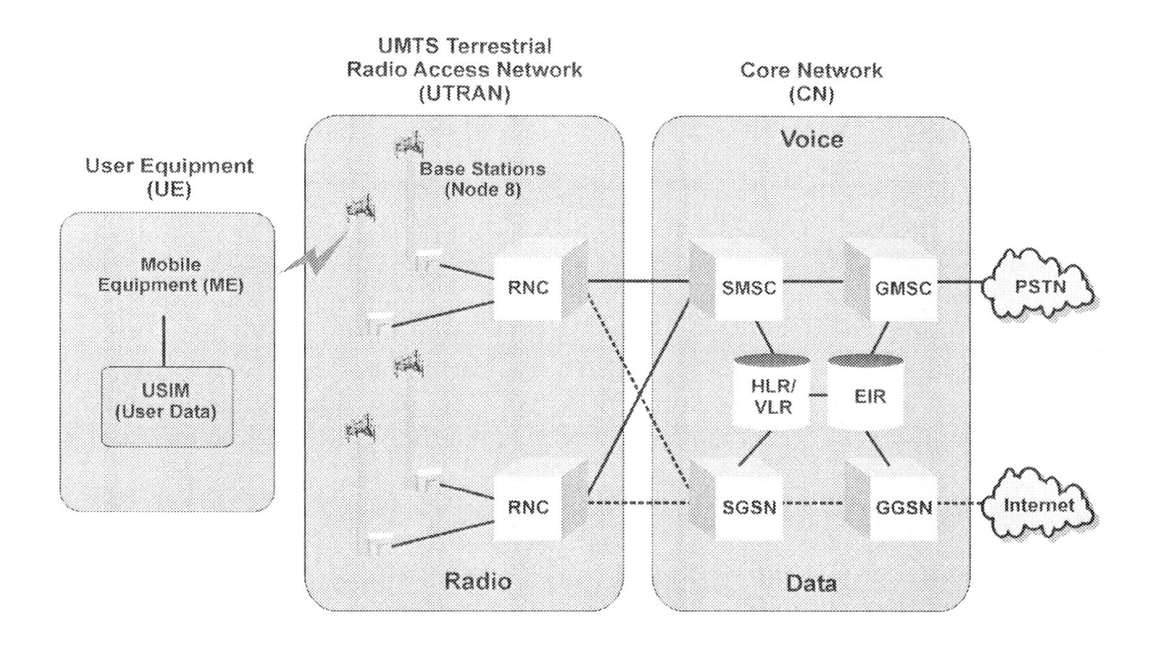

Figure 1.27., WCDMA Network

Base Stations (Node B)

Base stations (called Node B) may be stand alone transmission systems or part of a cell site and is composed of an antenna system (typically a radio tower), building, and base station radio equipment. Base station radio equipment consists of RF equipment (transceivers and antenna interface equipment), controllers, and power supplies. Base station transceivers have many of the same basic parts as a mobile device. However, base station radios are coordinated by the WCDMA system's BSC and have many additional functions than a mobile device.

The radio transceiver section is divided into transmitter and receiver assemblies. The transmitter section converts a voice or data signal to RF for transmission to wireless devices and the receiver section converts RF from the wireless device to voice or data signals routed to the MSC or packet switching network. The controller section commands insertion and extraction of signaling information.

Unlike end user wireless devices (such as a mobile telephone or laptop computer), the transmitter, receiver, and control sections of an access point may be grouped into equipment racks. For example, a single equipment rack may contain all of the RF amplifiers or voice channel cards. Unlike analog or early-version digital cellular systems that dedicated one transceiver in each base station for a control channel, the WCDMA system combines control channels and voice channels are mixed on a single physical radio channel.

Radio Antenna Towers

Wireless base station antenna heights can vary from a few feet to more than three hundred feet. Radio towers raise the height of antennas to provide greater area coverage. There may be several different antenna systems mounted on the same radio tower. These other antennas may be used for paging systems, a point to point microwave communication link, or land mobile radio (LMR) dispatch systems. Shared use of towers by different types of radio systems in this way is very common, due to the economies realized by sharing the cost of the tower and shelter. However, great care must be taken in the installation and testing to avoid mutual radio interference between the various systems.

Communication Links

Communication links carry both data and digital voice information between the base station and the WCDMA network. The physical connection options for communication links include copper wire, microwave radio, or fiber optic links. Duplicate and/or alternate communication links are sometimes used to help ensure communication may continue in the event of the failure of a communication line (such as when a cable seeking backhoe cuts a line).

Mobile Switching Center (MSC)

The mobile switching center (MSC) processes requests for service from mobile devices and land line callers, and routes calls between the base stations and the public switched telephone network (PSTN). The MSC receives the dialed digits, creates and interprets call processing tones, and routes the call paths.

Authentication, Authorization, and Accounting (AAA)

Authentication, Authorization, and Accounting are the processes used in validating the claimed identity of an end user or a device, such as a host, server, switch, or router in a communication network. Authorization is the act of granting access rights to a user, groups of users, system, or a process. Accounting is the method to establish who, or what, performed a certain action, such as tracking user connection and logging system users.

Interworking Function (IWF)

Interworking functions are systems and/or processes that attach to a communications network that is used to process and adapt information between dissimilar types of network systems. IWFs in the WCDMA system may include data gateways that convert circuit switched data from the MSC to the Internet.

Message Center (MC)

The message center is a node or network function within a communications network that accommodates messages sent and received via short messaging service (SMS).

Serving General Packet Radio Service Support Node (SGSN)

A serving general packet radio service support node is a switching node that coordinates the operation of packet radios that are operating within its service coverage range. The SGSN operates in a similar process of a MSC and a VLR, except the SGSN performs packet switching instead of circuit switching. The SGSN registers and maintains a list of active packet data radios in its network and coordinates the packet transfer between the mobile radios.

Gateway GPRS Support Node (GGSN)

A gateway GPRS support node is a packet switching system that is used to connect a GPRS packet data communication network to other packet networks such as the Internet.

Radio Network Controller (RNC)

A radio network controller is an automatic coordinator (controller) in a WCDMA system that allows one or more node B base transceiver station (BTS) to communicate with a mobile switching center and/or a packet data communication system. The RNC contains more control features than a traditional base station controller (BSC).

Voice Message System (VMS)

The voice mail system is a telecommunications system that allows a subscriber to receive and play back messages from a remote location (such as a PBX telephone or mobile phone). The VMS consists primarily of memory storage (for messages), telephone interfaces (to connect to the communication system), and message recording, playback, and control features (typically via DTMF tones).

Public Switched Telephone Network (PSTN)

Public switched telephone networks are communication systems that are available for public to allow users to interconnect communication devices. Public telephone networks within countries and regions are standard integrated systems of transmission and switching facilities, signaling processors, and associated operations support systems that allow communication devices to communicate with each other when they operate.

Public Packet Data Network (PPDN)

A packet data network that is generally available for commercial users (the public). An example of a PPDN is the Internet.

Network Databases

Network databases are information storage and retrieval systems that are accessible by a network. There are many network databases in the WCDMA network. Some of the key network databases include a master subscriber database (home location register), temporary active user subscriber database (visitor location register), unauthorized or suspect user database (equipment identity register), billing database, and authorization and validation center (authentication).

Home Location Register (HLR)

The home location register (HLR) is a subscriber database containing each customer's international mobile subscriber identity (IMSI) and international mobile equipment identifier (IMEI) to uniquely identify each customer. There is usually only one HLR for each carrier even though each carrier may have many MSCs.

The HLR holds each customer's user profile includes the selected long distance carrier, calling restrictions, service fee charge rates, and other selected network options. The subscriber can change and store the changes for some feature options in the HLR (such as call forwarding.) The MSC system controller uses this information to authorize system access and process individual call billing.

The HLR is a magnetic storage device for a computer (commonly called a hard disk). Subscriber databases are critical, so they are usually regularly backed up, typically on tape or CDROM, to restore the information if the HLR system fails.

Visitor Location Register (VLR)

The visitor location register (VLR) contains a subset of a subscriber's HLR information for use while a mobile telephone is active on a particular MSC. The VLR holds both visiting and home customer's information. The VLR eliminates the need for the MSC to continually check with the mobile telephone's HLR each time access is attempted. The user's required HLR information is temporarily stored in the VLR memory, and then erased either when the wireless telephone registers with another MSC or in another system or after a specified period of inactivity.

Equipment Identity Register (EIR)

The equipment identity register is a database that contains the identity of telecommunications devices (such as wireless telephones) and the status of these devices in the network (such as authorized or not-authorized). The

EIR is primarily used to identify wireless telephones that may have been stolen or have questionable usage patterns that may indicate fraudulent use. The EIR has three types of lists; white, black and gray. The white list holds known good mobile devices. The black list holds invalid (barred) mobile device. The gray list holds mobile devices that may be suspect for fraud or are being tested for validation.

Billing Center (BC)

A separate database, called the billing center, keeps records on billing. The billing center receives individual call records from MSCs and other network equipment. The switching records (connection and data transfer records) are converted into call detail records (CDRs) that hold the time, type of service, connection points, and other details about the network usage that is associated with a specific user identification code. These billing records are then transferred via tape or data link to a separate computer typically by electronic data interchange (EDI) to a billing system or company that can settle bills between different service providers (a clearinghouse company).

Authentication Centre (AuC)

The Authentication Centre stores and processes information that is required to validate ("authenticate") the identity of a wireless device before service is provided. During the authentication procedure, the AuC provides information to the system to allow it to validate the mobile device.

Number Portability Database (NPDB)

Number portability is the ability for a telephone number to be transferred between different service providers. This allows customers to change service providers without having to change telephone numbers. Number portability involves three key elements: local number portability, service portability and geographic portability. To enable number portability, the WCDMA system maintains a number portability database (NPDB). This database helps to route calls to their destination which may have an assigned telephone number that is different (number has been ported) than the destination phone number.

IP Backbone Network

A backbone network is the core infrastructure of a network that connects several major network components together. A backbone system is usually a high-speed communications network such as ATM or FDDI. The WCDMA system uses a backbone network that can provide end-to-end IP transmission capability.

Backbone network is a communications network that connects the primary switches or nodes within the network. The backbone network is usually composed of high-speed switches and communication lines.

The focus on using IP communication allows carriers to use off-the-shelf IP network equipment. This typically lowers the equipment cost (due to a large selection of vendors and equipment options), reduces operation and maintenance costs due to one type of system to maintain (less training and processes), and allows for the use of standard software (traffic monitoring and management).

Device Addressing

WCDMA mobile terminals are uniquely identified by the mobile identification number through the use of mobile station ISDN (MSISDN), of International Mobile Subscriber Identity (IMSI), International Mobile Equipment Identifier (IMEI), and a temporary mobile station identity (TMSI). In addition may be assigned temporary IP addresses as required.

Mobile Station ISDN (MSISDN)

The mobile station ISDN is the phone number assigned to mobile telephones. This number is compatible with the E.164 international public telephone numbering plan.

International Mobile Subscriber Identity (IMSI)

The international mobile subscriber identity is an identification number that is assigned by a mobile system operator to uniquely identify a mobile telephone.

International Mobile Equipment Identifier (IMEI)

An International Mobile Equipment Identifier (IMEI) is an electronic serial number that is contained in a GSM mobile radio. The IMEI is composed of 14 digits. six digits are used for the type approval code (TAC), two digits are used for the final assembly code (FAC), six digits are used for the serial number and two digits are used for the software version number.

Temporary Mobile Station Identity (TMSI)

A temporary mobile station identity (TMSI) is a number that is used to temporarily identify a mobile device that is operating in a local system. A TMSI is typically assigned to a mobile device by the system during its' initial registration. The TMSI is used instead of the International Mobile Subscriber Identity (IMSI) or the mobile directory number (MDN). TMSIs may be used to provide increased privacy (keeping the telephone number private) and to reduce the number of bits that are sent on the paging channel (the number of bits for a TMSI are much lower than the number of bits that represent an IMSI or MDN).

IP Address

Internet protocol addressing (IP addressing) is the use of unique identifiers in a data packet that are assigned to a particular device or portion of a device (such as a port) within a system or a domain (portion of a system). IP addressing varies based on the version of Internet protocol. For IP version 4, this is a 32-bit address and for IP version 6, this is a 128 bit address. To help simplify the presentation of IPv4 addresses, it is common to group each

8 bit part of the IP address with a decimal number separated from other parts by a dot(.), such as: 207.169.222.45. For IPv6 it is customary to represent the address as eight, four digit hexadecimal numbers separated by colons, such as 1234:5678:9000:0D0D:0000:5678:9ABC:8777.

While the WCDMA system was not designed to directly use IP addressing, IP addresses can be assigned to mobile devices when they are access data networks (such as the Internet).

The WCDMA system permits the static or dynamic assignment of IP addresses. Static IP addressing can simplify the connection of services to mobile devices. Dynamic IP addressing can better manage a limited number of IP addresses and enhance the security of systems.

Static IP Addressing

Static IP addressing is the process of assigning a fixed Internet Protocol (IP) address to a computer or data network device. Use of a static IP address allows other computers to initiate data transmission (such as a video conference call) to a specific recipient.

Dynamic IP Addressing

Dynamic IP addressing is a process of assigning an Internet Protocol address to a client (usually and end user's computer) on an as needed basis. Dynamic addressing is used to conserve on the number of IP addresses required by a server and to provide an enhanced level of security (no predefined address to use for hackers).

Figure 1.28 shows how a computer uses DHCP to obtain a temporary IP address when it requires an Internet communication session. In this example, the computer requests a connection with an Internet service provider (ISP) via a modem that is connected to a universal serial bus (USB) line. When the Internet service provider receives the request for connection, it assigns an IP address from is list of available IP addresses. The computer will then use this IP address for all of its communications with the Internet until it disconnects the connection to the ISP.

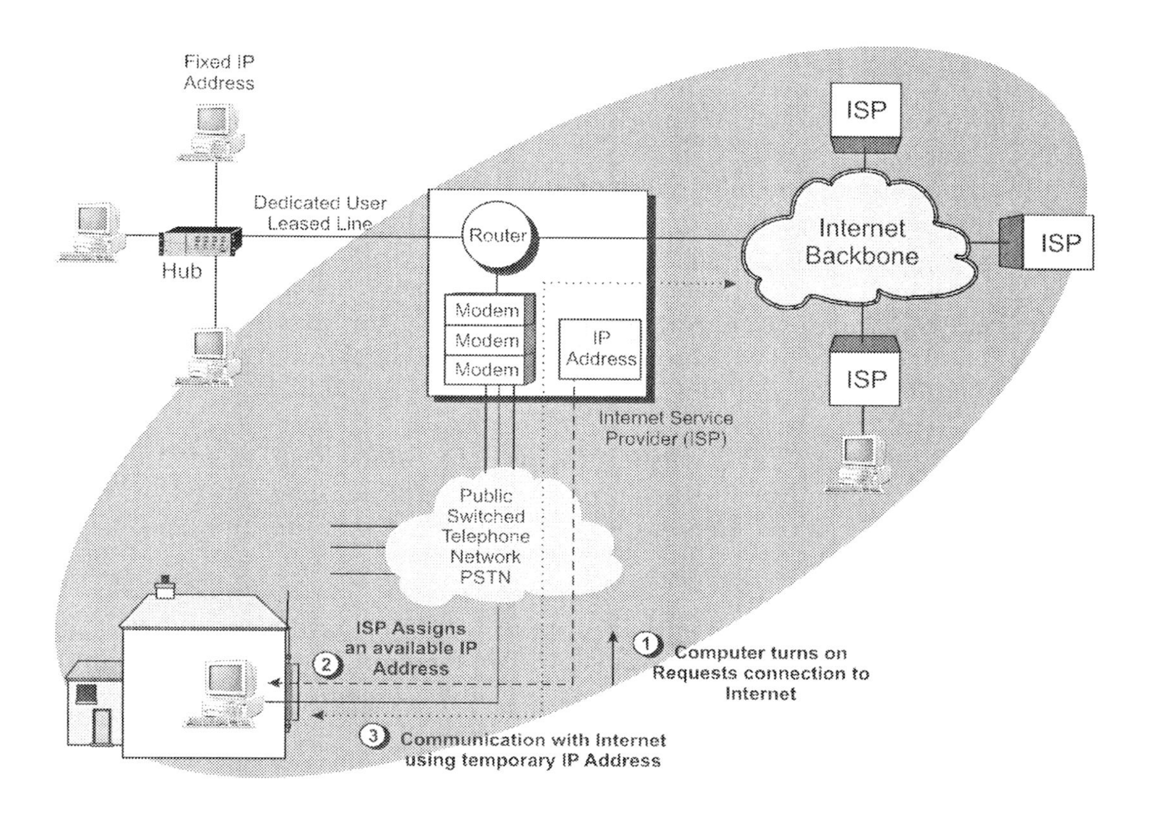

Figure 1.28., DHCP Addressing

WCDMA System Operation

The WCDMA system operation is the set of tasks performed to complete key operations: initialization of information when the subscriber unit is turned on, monitoring of control information, accessing the system, and maintaining communication sessions.

When a WCDMA mobile device is first turned on, it gathers its initial information (initializes) by scanning the available radio channels to find pilot channels. If it finds pilot signals, it determines that there is WCDMA service capability and it will begin to synchronize (time align) with the channel and obtain system broadcast information.

If it finds more than one pilot channel, it will usually tune to the radio channel with the strongest or best quality signal. If the mobile device does not find any pilot channels, it may begin to scan for GSM channels if the user has programmed the mobile device to allow access to the GSM system (companies may charge a surcharge for accessing GSM service.) If it finds GSM control channels, it will synchronize with the control channel and gather its system broadcast information.

The system broadcast information provides the information needed by the mobile device to monitor for incoming calls (paging messages) and to coordinate how to access the system (power level and maximum number of system access attempts).

After the mobile device has gathered its initial information (initialized), it will typically register with the system. This allows the system to route incoming calls to the cell site(s) that are able to communicate with the mobile device. The mobile device will continually register (sending short messages) to the system as it moves through new radio coverage areas.

When a WCDMA mobile device desires to obtain service from a mobile system, it sends an access request message. Before attempting to access the system, the mobile device must first listen to the control channel to determine if the system is idle (not currently busy) serving access requests from other mobile devices.

When a call is received by the wireless system for the mobile device, the system sends a call alert (page) message to the radio coverage area(s) where the mobile device has last registered. After the mobile device has initialized with the system, it constantly listens to the paging channels for page messages (its own identification number).

After the mobile device has gained the attention of the system and has been granted access to services, a communication session (connected mode) is established. During the communication session, a voice path and/or data paths may be opened (communication sessions). While the mobile device is in connected mode, other processes may simultaneously happen such as channel handover.

Initialization

Initialization is the process of a mobile device initially finding an available WCDMA radio channel, synchronizing with the system and obtaining system parameters from the pilot channel, synchronization channel (SCH), and broadcast channel (BCH) to determine the information requirements for access and communication.

Initialization phase begins when the access terminal first powers on. It initially looks to the UMTS subscriber identity module (USIM) card for a preferred control channel list. If there is no list, the mobile device scans all of the available radio channels to find a control channel.

Idle

During idle mode, the mobile device monitors several different control channels to acquire system access parameters, to determine if it has been paged or received an order, or to initiate a call (if the user is placing a call) or to start a data session (if the user has started a data application).

After obtaining the system parameters, the mobile device continuously monitors the broadcast channel for changes in system parameters, including system identification and access information. If the mobile device has discontinuous reception (sleep mode) capability, and if the system supports it, the mobile device turns off its receiver and other non-essential circuitry for a fixed number of burst periods. The system knows that it has commanded the mobile device to sleep, so it does not send pages designated for that mobile device during the sleep period. Because control channels are on only one of the 15 time slots in a frame, the mobile device can scan neighboring control channels during the unused time slots. If a better control channel (higher signal strength or better bit error rate) is available, the mobile device tunes to it.

The mobile device then monitors the paging control channel to determine if it has received a page. If a call is to be received, an internal flag is set indicating that the mobile device is entering access mode in response to a page. If the system sends an order such as a registration message, an internal flag is set indicating that the mobile device is attempting access in response to an order. When a user initiates a call, an internal flag is set indicating that the access attempt is a call origination, and dialed digits will follow the access request.

Access Control and Initial Assignment

Access control and initial assignment is the process of gaining the attention of the system, obtaining authorization to use system services, and the initial assignment to communication channel to setup a communication session.

Access control and initial assignment occurs when a mobile device responds to a page (incoming connection request), desires to setup a call, or any attempt by the mobile device. Access to the WCDMA is a random (at any time) occurrence (not usually preplanned.) To avoid access "collisions" between mobile devices, a seizure collision avoidance process is used. Before a mobile device attempts access to the system, it first waits until the channel is available (not busy serving other users). The mobile device then begins transmitting an access request message (called an access probe) on the Random Access Channel (RACH) at very low power. An access probe contains a preamble that is followed by the access channel data message. The access probe preamble uses a predefined sequence and a fixed spreading that allows the system to easily detect an access probe message. The access request indicates the type of service the mobile device is requesting (e.g. call origination or a response to a paging message).

If an access request message does not gain the attention of the system in a short period of time, the mobile device will increase its transmitter power level and send another access probe. This process will repeat until either the system responds to the access request on the AICH channel or if the mobile device reaches a maximum allowable transmission power level established by the system.

If the system acknowledges the mobile device's request for service, the mobile device will send additional information to the system that allows it to setup a dedicated communications channel where conversation or data transmission can begin.

Figure 1.29 shows the process used by mobile devices to gain access to a WCDMA system. This diagram shows that the mobile device transmits an access probe packet to the system to attempt to gain the attention of the sys-

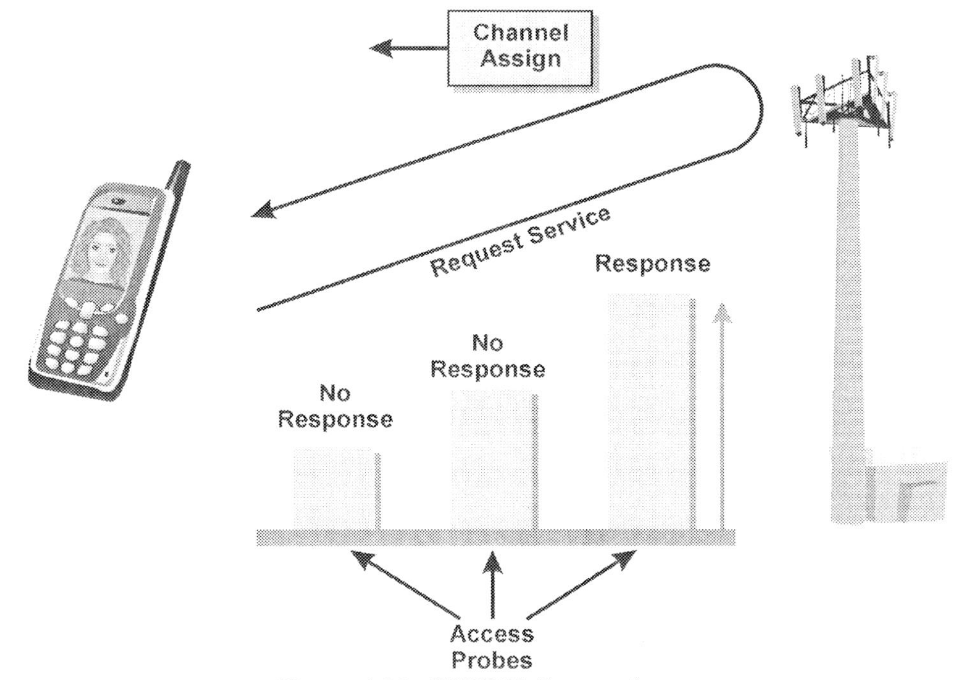

Figure 1.29., WCDMA System Access

tem. This example shows that the mobile device transmits the first access probe at low power. If the system does not respond within a short amount of time, the mobile device will transmit another access probe at a higher RF power level. This process of sending an access probe at a higher power level and waiting for the system to respond continues until either the system responds to the access probe request or the mobile device reaches its maximum access probe RF power level that is assigned by the system.

Authentication

Authentication is a process of exchanging information between a communications device (typically a user device such as a mobile phone) and a communications network that allows the carrier or network operator to confirm the true identity of the user (or device). This validation of the authenticity of the user or device allows a service provider to deny service to users that cannot be identified. Thus, authentication inhibits fraudulent use of a communication device that does not contain the proper identification information. The WCDMA may require the mobile device to authenticate with the system during the system access process.

Paging

Paging is a process used to alert mobile devices that they are receiving a call, command, or message. Mobile devices listen for paging messages for their identification code (IMSI number or TMSI) on a paging channel.

After a mobile device has registered with the system, it is assigned to a paging group. The paging group is identified by a paging indicator (PI) that is provided at the beginning of the paging message group. The mobile device first reads the PI to determine if it should remain awake to receive the paging group or if it can go to sleep as its identification code is not associated with the particular paging group.

Connected Mode

Connected mode is the process of managing the communication session when a mobile device is transferring data to and from a Base Station. When in the connected mode, the base transceiver station (BTS) continuously controls the mobile device during the communication session. These control tasks include power level control, hand-over, alerting, etc. The Base Station exercises control during the communication session through dedicated control channels.

To enter the connected mode, the base station must open a communication channel with the mobile device. When the connection is opened, each frame or packet that is received by the base station can be transferred to the assigned communication line (for a voice call) or IP address (for a data communications session). When a communication session is complete (e.g. the user presses end or closes their email or web browsing application), the connection is closed and the base station may assign other users to the radio resources.

The mobile device does not have to continuously transmit data while in the connected mode. When in the connected mode, the base station associates (maps) the radio link to the circuit switched communication channel (for voice) or to an IP address (for data). When the connected mode is used for data transmission such as web browsing, the typical data transmission activity is less than 10%. Other mobile devices can use the channels during inactive data transmission periods allowing a system to serve many (hundreds) simultaneous data users for each WCDMA radio channel.

During the connected mode, communication session processing tasks include the insertion and extraction of control messages that allow functions such as power control monitoring and control, handover operation, adding or terminating additional communication sessions (logical channels), and other mobile device operational functions.

Packet Data Scheduling Algorithm

A packet data scheduling algorithm is a program that coordinates the sequences of processes or information. The packet data scheduling algorithm in the WCDMA system is used to coordinate the flow of data to multiple WCDMA users.

Mobile data usage such as Internet web browsing involves data transmission that is not continuous. The ratio of data transmitted by a single device compared to the overall data transmission by all the devices is the activity factor. The lower the activity factor, the higher the number of mobile data devices that can access the system.

Packet scheduling can assign different priority levels based on user and application types. Packets for specific types of users such as public safety officers can be given higher priority. Packets for specific applications such as IP Telephony can be given priority over other applications such as web browsing or email access.

Registration

Registration is the process that is used by mobile devices to inform the wireless system of their location and availability to receive communications services (such as incoming calls). The reception of registration requests allows a wireless system to route incoming messages to the radio base station or transmitter where the mobile device has recently registered.

The process of registration is typically continuous. Mobile devices register when they power on, when they move between new radio coverage areas, when requested by the system, and when the mobile device is power off.

Because the registration process consumes resources of the system (channel capacity and system servicing capacity), there is a tradeoff between regularly maintaining registration information and the capacity of the system. During periods of high system usage activity, registration processes may be reduced.

Radio Signaling Protocols

Signaling is a process of transferring control information such as address, call supervision, or other connection information between communication equipment and other equipment or systems.

The WCDMA system was designed to use multiple protocols that can be divided into processing layers. Each layer in the protocol performs specific functions and each layer may use one or more protocols. The use of layers allows for the addition of functions and modification of protocols without requiring significant changes to the system.

Different parts of the UMTS system such as the radio interface, radio access network, and intersystem communication use different protocols and protocol layers. The WCDMA radio interface system protocol layers include radio resource control (RRC), radio link control (RLC), medium access control (MAC) layer, and physical layer.

Radio Resource Control (RRC)

The radio resource control (RRC) is a layer 3 (network) protocol that controls the setup, management, and termination of physical and logical channels between the base station and the mobile device. It oversees the signaling on the common control and dedicated control channels. RRC signaling messages are also used to provide for channel quality measurements that are used for channel handovers.

Packet Data Convergence Protocol (PDCP)

The packet data convergence protocol (PDCP) coordinates the efficient transfer and control of packet data transmission. The main functions of PDCP include compression of the headers during packet transmission over the radio channel (remove redundant packet header information) and to ensure reliable packet transfer (sequentially numbering and verifying packet delivery.)

Broadcast and Multicast Control Protocol (BMC)

Broadcast and multicast control (BMC) protocol coordinates the transmission and reception of broadcast and multicast services. The main functions of BMC include the reception, storage, and distribution of broadcast and multicast messages such as broadcast short messages to the appropriate applications.

Radio Link Control (RLC)

Radio link protocol (RLP) is a layer 2 (link layer) that is used to coordinate the overall flow of data packets across the radio link. RLP uses error detection and data retransmission to increase the reliability of the radio link while reducing the error rate. RLC functions include packet division (segmentation), reassembly, concatenation, padding, and error correction when the communication channel requires error control (not all sessions require error detection and control). The RLC is also used to control the flow of data (rate of transfer).

Medium Access Control (MAC) Layer

The medium access control (MAC) layer is used to coordinate access between mobile devices and base stations. The MAC layer is composed of one or more communication channels that are used to coordinate the access of communication devices to a shared communications medium or channel (the radio link). MAC channels typically communicate the availability and access priority schedules for devices that may want to gain access to a communication system.

The MAC layer is also used to coordinate the data transmission rate on the traffic channels and to assign (map) logical channels to underlying physical RF channels. After the access terminal has competed to gain the attention of the system and it has been assigned a connection, the MAC layer is used to coordinate the sharing of (competition for) the traffic channel for packet data transfer.

Physical Layer

The physical layer performs the conversion of data to a physical medium (radio) transmission and coordinates the transmission and reception of these physical signals. The physical layer receives data for transmission from upper layers (the MAC layer) and converts it into physical format suitable for transmission to a device or through the network.

Figure 1.30 shows the protocol layers of the WCDMA radio interface. This diagram shows that each layer in the WCDMA radio link has its own function and as data passes through each network layer, the layer performs its function and passes its data on to the next layer above or below. The radio resource control (RRC) protocol is used to coordinate the operation (control) of the radio. The broadcast and multicast control (BMC) protocol layer is responsible for receiving and processing broadcast messages. The packet data convergence protocol (PDCP) layer is concerned with ensuring all the packets are transferred and placed in correct order. The radio link control layer is concerned with maintaining the radio link between the mobile device and the base station. The medium access control layer (MAC) is the process used to request and coordinate access to the system. The physical layer converts digital bits to and from RF packets that are sent between the WCDMA device and the access point.

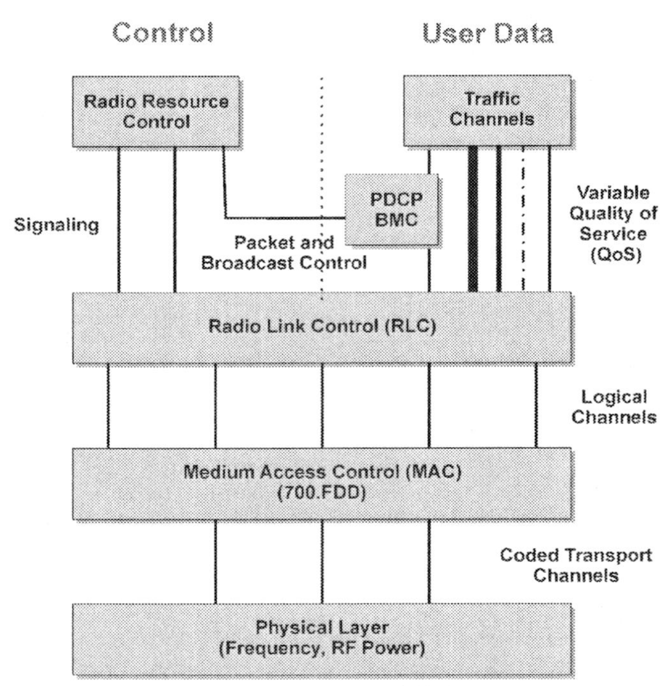

Figure 1.30., WCDMA Radio Link Protocol Layers

WCDMA Multimedia Signaling Protocols

The WCDMA system can use upper layer signaling protocols to provide for advanced multimedia services and features. Some of the protocols used with the WCDMA system include session initiation protocol (SIP), session description protocol (SDP), and real time signaling protocol (RTP).

Session Initiation Protocol (SIP)

SIP is an application layer protocol that uses text format messages to setup, manage, and terminate multimedia communication sessions. SIP is a simplified version of the ITU H.323 packet multimedia system. SIP is defined in RFC 2543.

Session Description Protocol (SDP)

Session description protocol (SDP) is a text based protocol that is used to provide high-level definitions of connections and media streams. The SDP protocol is used with session initiated protocol (SIP). The SDP protocol is used in a variety of communication systems including 3G wireless and the PacketCable system. SDP is defined in RFC 2327.

Real Time Transport Protocol (RTP)

RTP is a packet based communication protocol that adds timing and sequence information to each packet to allow the reassembly of packets to reproduce real time audio and video information. RTP is defined in RFC 1889. RTP is the transport used in IP audio and video environments.

WCDMA Future Evolution

The future evolution of WCDMA is likely to include increased data transmission rates that are achieved by multichannel carrier system, increased modulation efficiency, and directional antenna systems.

Increased Data Transmission Rates

The CDMA system has evolved and continues to evolve to provide higher data transmission rates in both the downlink (system to the CDMA device) and in the uplink (from the CDMA device back to the system). The data transmission rate should continue to increase through the use of more efficient modulation technologies and increased radio channel bandwidth.

Modulation and radio spectrum efficiency are measurements characterize by how much information can be transferred in a given amount of frequency bandwidth. This is often given as bits per second per Hertz. The use of more efficient modulation and coding methods that have high spectral efficiency also typically are very sensitive to small amounts of noise and interference, and often have low geographic spectral efficiency. (See also Geographic Spectral Efficiency)

Multichannel Carrier

Increased radio channel bandwidth can be accomplished by combining multiple carriers (multichannel carrier) and/or by defining a new wide radio channel. Multichannel carrier (MC) is a communication system that combines or binds together two or more communication carrier signals (carrier channels) to produce a single communication channel. This single communication channel has capabilities beyond any of the individual carriers that have been combined. The use of a wider radio channel would allow significant increases in data transmission rates but it would require new mobile and system equipment and probably would require new frequency or transmission planning.

Figure 1.31 shows how a wireless communication system can combine multiple radio communication channels to provide a communication channel that has higher data transmission rates. This diagram shows a wireless communication system that contains multiple radio channels. This example shows that an incoming 1 Mbps data signal has a higher data transmission rate than a single radio channel can provide. To transfer the high-speed data signal, is it split into two (or more) data channels with a lower data transmission rate. Each lower-speed data channel is then sent to an RF channel transmitter. To coordinate the overall flow of data, an additional control function is needed that coordinates the flow of data on the RF channel and to insert multicarrier control messages that allow the receiver of the multiple channels to know how the channels are combined. This diagram shows that the mobile device is able to simultaneously receive and combine each of the radio channels to produce the original high-speed digital signal.

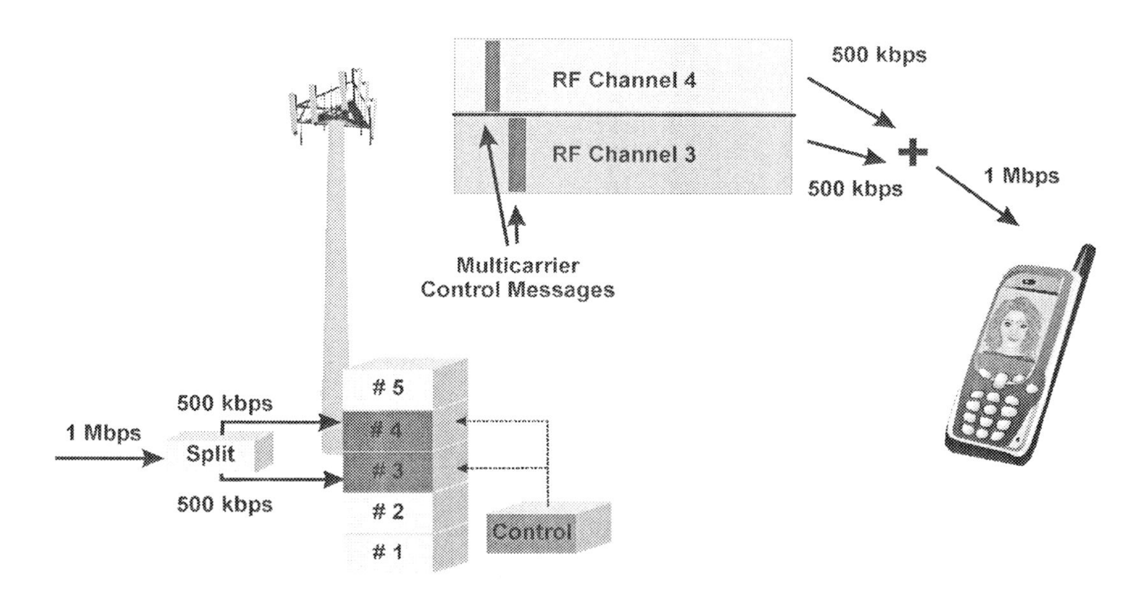

Figure 1.31., Multichannel Carrier

More Efficient Modulation Technologies

Modulation is the process of changing the amplitude, frequency, or phase of a radio frequency carrier signal (a carrier) to change with the information signal (such as voice or data). Modulation efficiency is a measure of how much information can be transferred onto a carrier signal. In general, more efficient modulation processes require smaller changes in the characteristics of a carrier signal (amplitude, frequency, or phase) to represent the information signal.

To increase the data transmission rates, more efficient (and more complex) modulation technologies can be used. This includes combining multiple modulation types. The WCDMA system is evolving to use more efficient modulation technologies such as quadrature amplitude modulation (QAM) to allow each symbol (carrier change) to represent more information bits.

Spatial Division Multiple Access (SDMA)

Spatial division multiple access (SDMA) is a system access technology that allows a single transmitter location to provide multiple communication channels by dividing the radio coverage into focused radio beams that reuse the same frequency. To allow multiple access, each mobile radio is assigned to a focused radio beam. These radio beams may dynamically change with the location of the mobile radio. SDMA technology has been successfully used in satellite communications for several years.

This figure shows how an IP TV system uses relatively simple text messages to setup and control television channels. This diagram shows how a television has IP TV capability via a set top box that is controlled by an IP TV server. This IP TV based television set top box is called a User Agent (UA). The User Agent (UA) is actually a gateway that converts video, audio, and control information (e.g. channel change requests) into packets that can be routed through a data network (such as the Internet) to call servers and other User Agents (UAs.) The control packets are sent to and from the call server to request and receive television channels. IP TV servers may communicate with other call servers to setup distant television channel connections. This diagram shows how a distant IP TV server controls a User Agent (UA) gateway that allows television channels to connect via Internet to various IP TV media providers.

The WCDMA system was designed to allow the use of SDMA technology. A secondary pilot channel was included in the system design to assist in the selection and focusing of directional antennas.

Figure 1.32 shows an example of an SDMA system. Diagram (a) shows the conventional sectored method for communicating from a cell site to a mobile telephone. This system transmits a specific frequency to a defined (sectored) geographic area. Diagram (b) shows a top view of a cell site that uses SDMA technology that is communicating with multiple mobile telephones operating within the same geographic area on a single frequency. In the SDMA system, multiple directional antennas or a phased array antenna system directs independent radio beams to different directions. As the mobile telephone moves within the sector, the system either switches to an alternate beam (for a multi-beam system) or adjusts the beam to the new direction (in an adaptive system).

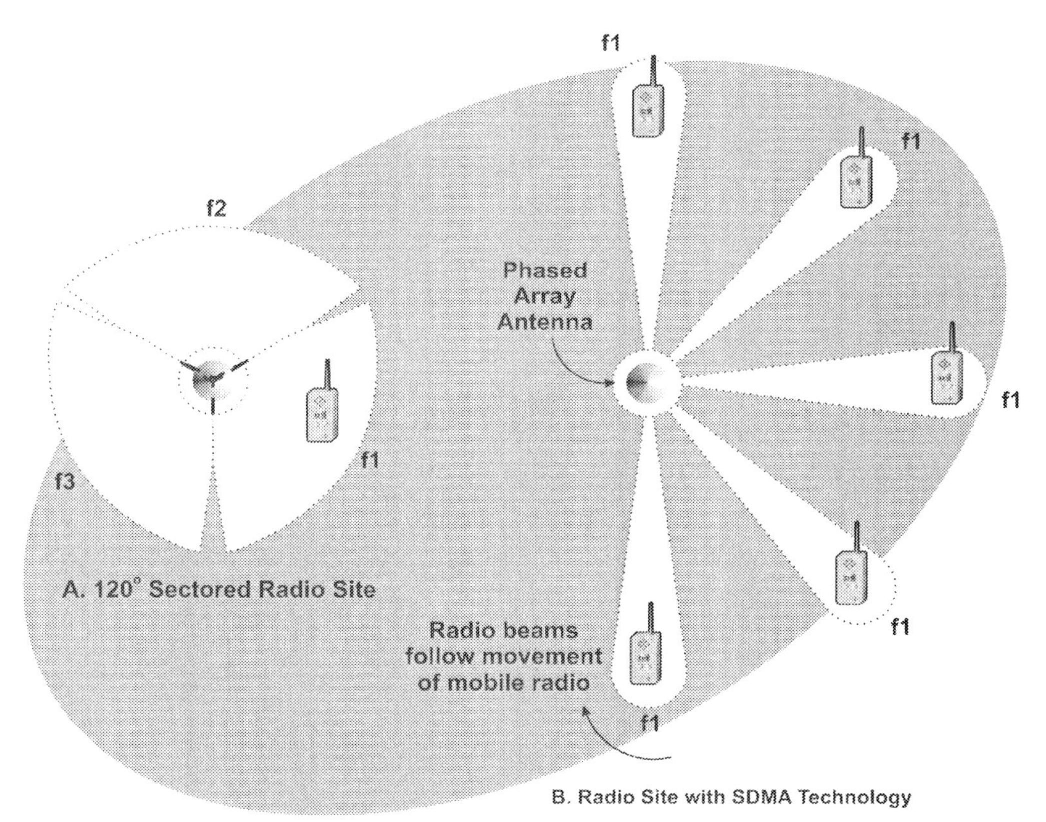

Figure 1.32., Spatial Division Multiple Access

Index

Wireless Books

by ALTHOS Publishing

Wireless Systems

ISBN: 0-9728053-4-6 Price: $34.99
Authors: Lawrence Harte, Dave Bowler, Avi Ofrane, Ben Levitan #Pages 368 Copyright Year: 2004

Wireless Systems; Cellular, PCS, 3G Wireless, LMR, Paging, Mobile Data, WLAN, and Satellite explains how wireless telecommunications systems and services work. There are many different types of wireless systems competing to offer similar types of voice, data, and multimedia services. This book describes what the functional parts of these systems are and the basics of how these systems operate. With this knowledge, you can understand what types of services can these systems effectively offer and how they compare to competing systems. Covered in this book are the key market segments and trends for these systems along with a description of the leading commercial systems and the services they offer. Also included in the book are descriptions of the common services that are offered by each system and how and the amount this that typically charged to the consumers for these services.

Wireless Dictionary

ISBN: 0-9746943-1-2 Price: $39.99
Author: Althos #Pages: 628 Copyright Year: 2004

The Wireless Dictionary is the Leading wireless industry resource. The Wireless Dictionary provides definitions and illustrations covering the latest voice, data, and multimedia services and provides the understanding needed to communicate with others in the wireless industry. This book is the perfect solution for those involved or interested in the operation of wireless devices, networks, and service providers. This reference book explains the latest technologies and applications used in the wireless industry, assists with the explanation of technologies by using many diagrams and pictures. It is a great reference tool that allows people to effectively communicate with other people involved in the wireless industry. The convergence of technologies and systems means more competitors and new industry terms. As a result, communicating with others has become an alphabet soup of acronyms and technical terms.

Introduction to 802.11 Wireless LAN (WLAN)

ISBN: 0-9746943-4-7 Price: $14.99
Author: Lawrence Harte #Pages: 52 Copyright Year: 2004

Introduction to 802.11 Wireless LAN (WLAN), Technology, Installation, Setup, and Security book explains the functional parts of a Wireless LAN system and their basic operation. You will learn how WLANs can use access points to connect to each other or how they can directly connect between two computers. Explained is the basic operation of WLAN systems and how the performance may vary based on a variety of controllable and uncontrollable events. This book will explain the key differences between the WLAN system versions such as frequency and data rate along with which versions are compatible with each other. Tips are included on how to install a WLAN system along with common problems you may encounter and solutions you may use. Introduction to 802.11 Wireless LAN (WLAN) explains the basics of WLAN technology, how to install and setup a basic WLAN system, and key security options that should be considered.

Introduction To Wireless Systems

ISBN: 0-9742787-9-3 Price: $11.99
Author: Lawrence Harte, #Pages: 68 Copyright Year: 2003

Introduction to Wireless Systems book explains the different types of wireless technologies and systems, the basics of how they operate, the different types of wireless voice, data and broadcast services, key commercial systems, and typical revenues/costs of these services. Wireless technologies, systems, and services have dramatically changed over the past 5 years. New technology capabilities and limited restrictions (deregulation) are allowing existing systems to offer new services. Many of these new services compete with other types of wireless systems that have not experienced significant competition. While new competition has provided lower cost services for consumers, it means a rapidly changing marketplace for the wireless industry. Some of these changes include the increase in mobile telephone customers from 300 million to 1.2 billion customers within 5 years.

Introduction to Paging Systems

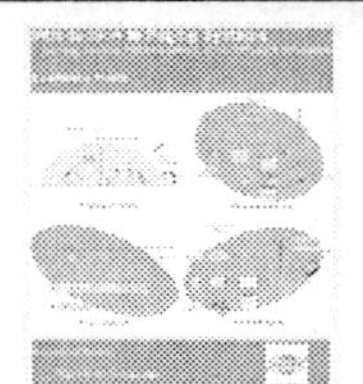

ISBN: 0-9746943-7-1 Price: $14.99
Author: Lawrence Harte #Pages: 48 Copyright Year: 2004

Introduction to Paging Systems describes the different types of paging systems, what services they can provide, an they are changing to meet new types of uses. This book explains the different types of paging systems and how th changing. Traditional paging services have seen a world decline of over 65% in the number of human between 1 2003. During this decline, some paging systems have experienced high-growth in new areas such as short mess and telemetry. Paging systems provide reliable information transfer that reaches many places that are not reach other types of mobile or satellite systems. Explained is how and why paging systems are transitioning from one-wa tems to two-way systems.also covered in this book are the key paging industry standards.

Introduction to Satellite Systems

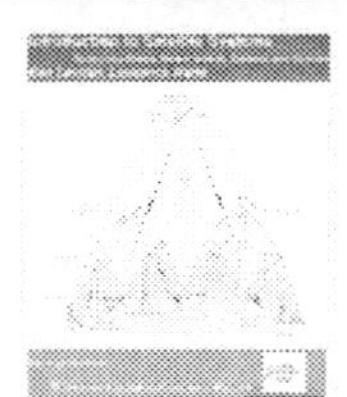

ISBN: 0-9742787-8-5 Price: $11.99
Author: Ben levitan, Lawrence Harte #Pages: 48 Copyright Year: 2004

In 2003, the satellite industry was a high-growth business that achieved over $83 billion in annual revenue. This offers an introduction to existing and soon to be released satellite communication technologies and services. It c how satellite systems are changing, growth in key satellite markets, key technologies that are used in satellite sys commercial satellite systems, and provides information on the leading services and their costs for satellite comm tion. Each type of satellite system and its technologies have unique advantages and limitations, which offer importan nomic and technical choices for managers, salespeople, technicians, and others involved with satellite equipme systems. This Introduction to Satellites book provides a basic understanding of the major satellite systems&technol

Introduction to Mobile Data

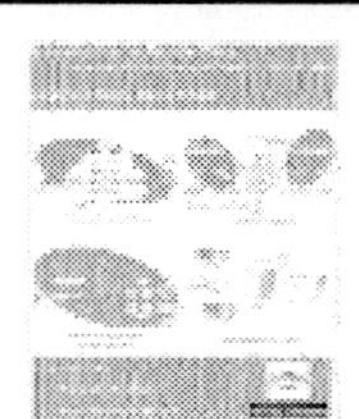

ISBN: 0-9746943-9-8 Price: $14.99
Author: Lawrence Harte #Pages: 628 Copyright Year: 2004

Introduction to Mobile Data explains how people use devices that can send data via wireless connections, what sy are available for providing mobile data service, and the services these systems can offer. This book explains the of circuit switched and packet data via wireless mobile systems. Included are descriptions of various public and p systems that are used for data and messaging services. This book explains the basics on how users send data th a variety of mobile communication systems. The mobile data marketplace has dramatically increased as a result mobile data applications, much lower cost of sending mobile data, and increased consumer awareness of the ava ty of mobile data services. Despite these dramatic improvements, this book describes how mobile data devices an tems still have reliability, cost, and bandwidth limitations compared to wired communication systems.

Introduction to Private Land Mobile Radio

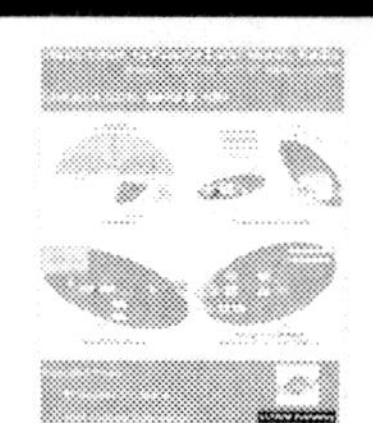

ISBN: 0-9746943-6-3 Price: $14.99
Author: Lawrence Harte #Pages: 50 Copyright Year: 2004

Introduction to Private Land Mobile Radio explains the different types of private land mobile radio systems, their operation, and the services they can provide. This book covers the basics of private land mobile radio systems i ing traditional dispatch, analog trunked radio, logic trunked radio (LTR), and advanced digital land mobile radio sys Described are the basics of LMR technologies including simplex, half-duplex, and full duplex operation. The dif types of squelch systems are covered including carrier controlled squelch, tone controlled squelch, and digital sq The basics of analog and digital trunked radio systems are provided along with how and why analog trunked radi tems are converting to digital trunked radio systems. The leading LMR industry standards including APCO, ED MPT1327, iDEN, and Tetra are described along with simple diagrams to explain their operation.

Introduction to GSM Systems

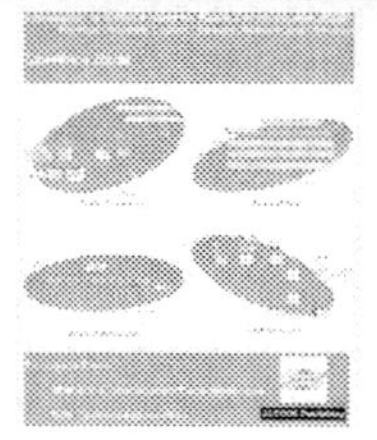

ISBN: 1-9328130-4-7 Price: $14.99
Author: Lawrence Harte #Pages: 48 Copyright Year: 2004

Introduction to GSM describes the fundamental components, key radio and logical channel structures, and the basic ation of the GSM system. This book explains the basic technical components and operation of GSM technology. Yo learn the physical radio channel structures of the GSM system along with the basic frame and slot structures. Desc are the logical channels and their functions along with the key GSM network components and how they commu with each other.

Internet Telephone Basics

ISBN: 0-9728053-03 Price: $29.99
Author: Lawrence Harte #Pages:226 Copyright Year: 2003

Internet Telephone Basics explains how and why people and companies are changing to Internet Telephone Service. Learn how much money can be saved using Internet telephone service and how you can to use standard telephones and dial the same way. Internet telephone service usually costs 1.5 to 4 cents/min for long distance calls and 3 to 10 cents/min for International calls. It describes how to activate Internet telephone service instantly and how to display your call details on the web. Covered are the advanced features and services including intelligent call forwarding, unified email and voice mail messaging, and the simultaneous sending of voice, data, and video through the Internet during your calls.

Voice Over Data Networks for Managers

ISBN: 0-9728053-2-X Price: $49.99
Author: Lawrence Harte #Pages:352 Copyright Year: 2003

Voice over Data Networks for Managers explains how to reduce communication costs 40% to 70%, keep existing telephone systems, and ways to increase revenue from new communication applications. Discover the critical steps companies should take and risks to avoid when transitioning from private telephone systems (KTS, PBX, and CTI) to provide voice over data (VoIP) services. Understand IP Centrex and Internet PBX (iPBX) systems and the different types of telephones, call servers and features they use. Learn how to get the necessary quality of service (QoS), security, and reliability you expect from traditional telephone systems.

Patent or Perish

ISBN: 0-9728053-3-8 Price: $39.95
Author: Eric Stasik #Pages:220 Copyright Year: 2003

Patent or Perish Explains in clear and simple terms the vital role patents play in enabling high technology firms to gain and maintain a competitive edge in the knowledge economy. Patent or Perish is a Guide for Gaining and Maintaining Competitive Advantage in the Knowledge Economy. In a world where knowledge has value and knowledge creates value, ideas are the new source of wealth. This book describes how technologies like the Internet remove traditional barriers to entry and enable competitors to quickly and effortlessly duplicate positions of competitive advantage and shows why just having good ideas is not enough. This books shows why companies need good ideas that can be patented.

Telecom Basics 3rd Edition

ISBN: 0-9728053-5-4 Price: $34.99
Author: Lawrence Harte #Pages:356 Copyright Year: 2003

This introductory book provides the fundamentals of signal processing, signaling control, can call processing technologies that are used in telecommunication systems. Covered are the key facets of voice and data communications, ranging from such basics as to how a telephone set works to more complex topics as how to send voice over data networks and the ways calls are processed in public and private telephone systems. The reader will learn about analog and digital signals, signal modulation, channel coding, and other types of signal processing. Copper wire, radio, and optical transmission systems are explained. Circuit and packet switching technologies are described along with the E&M line trunk signaling and SS7 control systems that are used to manage them. Explained or the different types of communication protocols such as IP, TCP, UDP and RTP, their stack layers, and how the perform parameter negotiation.

Introduction to Cable Television Systems

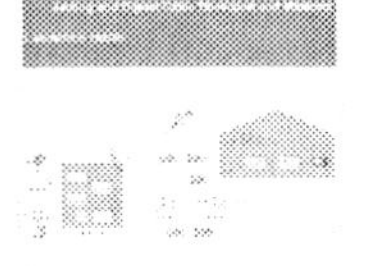

ISBN: 0-9728053-6-2 Price: $12.99
Author: Lawrence Harte #Pages: 48 Copyright Year: 2004

Community access television (CATV) is a television distribution system that uses a network of cables to deliver multiple video, data, and audio channels. This excerpted chapter from Telecom Systems provides an overview of cable television systems including cable modems, digital television, high definition television (HDTV), and the market growth of cable television and advanced services such as video on demand. Some of the topics described include: broadcast television, cable television, cable converters, distribution network, head-end, market growth, analog video, digital video, signal scrambling, National Television Standards Committee (NTSC), Phase Alternating Line (PAL), Sequential Couleur Avec Memoire (SECAM), Motion Picture Experts Group (MPEG) Compression, Data Over Cable Service Interface Specifications (DOCSIS), and Pay per View (PPV). Future enhancements such as wireless cable, cable telephony.

Introduction to Private Telephone Networks 2nd Edition

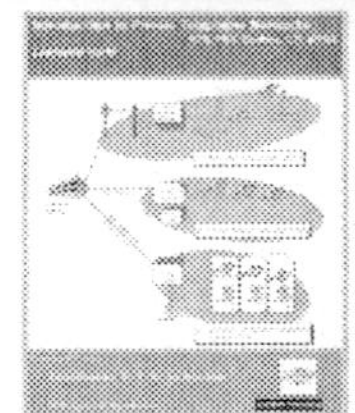

ISBN: 0-9742787-2-6 Price: $12.99
Author: Lawrence Harte #Pages: 48 Copyright Year: 2004

Private telephone networks are communication systems that are owned, leased or operated by the companies tha these systems. They primarily allow the interconnection of multiple telephones within the private network with each and provide for the sharing of telephone lines from a public telephone network. This excerpted chapter from Te Made Simple is an overview of independent telephone systems including private branch exchange (PBX) and con telephone integration. Included are descriptions of telephone stations, local wiring, switching systems, numbering market growth, digital packet voice telephony, Automatic Call Distribution (ACD), Interactive Voice Response (IVR Telephone Systems (KTS), Wireless Private Branch Exchange (WPBX), and direct marketing.

Introduction to Telecom Billing

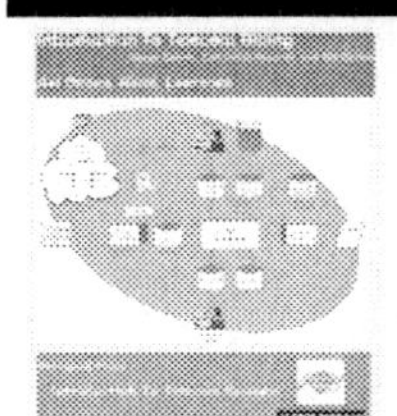

ISBN: 0-9742787-4-2 Price: $11.99
Author: Lawrence Harte #Pages: 36 Copyright Year: 2003

This book explains how companies bill for telephone and data services, information services, and non-communic products and services. Billing and customer care systems convert the bits and bytes of digital information within a work into the money that will be received by the service provider. To accomplish this, these systems provide accoun vation and tracking, service feature selection, selection of billing rates for specific calls, invoice creation, payment and management of communication with the customer. The authors have worked with hundreds of companies and types of billing system and discovered that in the early 2000s, the functions of billing systems were dramatically c ing due to the combining of voice, data and other types of services. Billing systems have also been transforming to for charging of non-traditional products and services such as candy from vending machines, tickets for entertair events, and home delivery of pizza.

Introduction to Public Switched Telephone Networks 2nd Edition

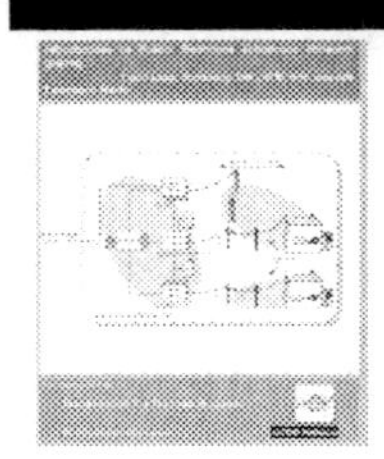

ISBN: 0-9742787-6-9 Price: $34.99
Author: Lawrence Harte #Pages: 48 Copyright Year: 2004

Public telephone networks are unrestricted dialing telephone networks that are available for public use to interco communications devices. There are also descriptions of many related topics, including: Local loops, switching sys numbering plans, market growth, public telephone system interconnections, common channel signaling (SS7), adva intelligent networks (AIN), plain old telephone service (POTS), integrated digital services network (ISDN), digita scriber line (DSL), digital loop carrier (DLC), passive optical network (PON), and public services. Future enhance such as high-speed multimedia services, packetized voice, fiber distribution networks, and soft switches are includ

Introduction to SS7 & IP Telephony

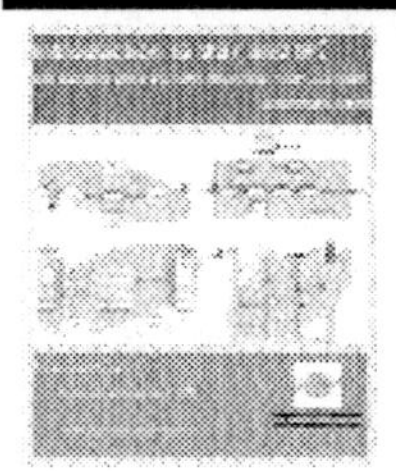

ISBN: 0-9746943-0-4 Price: $14.99
Author: Lawrence Harte #Pages: Copyright Year: 2004

The Introduction to Signaling System 7 (SS7) and IP control system that is used in public switched telephone net (PSTN) can be interconnected to other types of systems and networks using Internet Protocol (IP). Some of the connection issues relate to how the control of devices can be performed using dissimilar systems (e.g. mixing voic data systems). Another key reason for interconnecting SS7 network devices to IP data networks is the cost saving result from avoiding the access charges for connect equipment to SS7 systems and databases. This excerpted ch from SS7 Basics, 2nd Edition provides an overview of how SS7 and Internet Protocol (IP) can be integrated. It is an duction on how SS7 messages can be transported over IP networks (even the Internet in some cases).

Introduction to IP Telephony

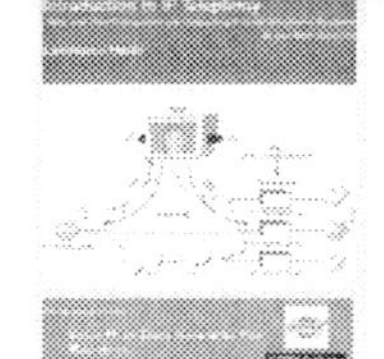

ISBN: 0-974278-7-7 Price: $12.99
Author: Lawrence Harte #Pages: Copyright Year: 2003

This "Introduction to IP Telephony" book explains why companies are converting some or all of their telephone sys from dedicated telephone systems (such as PBX) to more standard IP telephony systems. These conversions allo telephone bill cost reduction, increased ability to control telephone services, and the addition of new telephone infc tion services. By upgrading their systems, companies can immediately reduce their telecommunication costs 40% to Because IP telephony systems allow the end user and system administrators to setup and disconnect telephone bers and services, this provides increased control over their telephone features and services. IP telephony is us based on standard data formats (Internet Protocol).

Signaling System Seven (SS7) Basics 3rd Edition

ISBN: 0-9728053-7-0 Price: $34.99
Author: Lawrence Harte #Pages: 276 Copyright Year: 2003

This introductory book explains the operation of the signaling system 7 (SS7) and how it controls and interacts with public telephone networks and VoIP systems. SS7 is the standard communication system that is used to control public telephone networks. In addition to voice control, SS7 technology now offers advanced intelligent network features and it has recently been updated to include broadband control capabilities, local number portability, and mobile communication services. SS7 networks are now interconnecting with and operating on Internet data networks. This book will help the reader gain an understanding of SS7 technology, network equipment, and overall operation. It covers the reasons why SS7 exists and is necessary, as well as step-by-step procedures that describe the actions that occur in the network. SS7 Basics, 3rd Edition is for both the technical and the non-technical reader alike.

Introduction to SIP IP Telephony Systems

ISBN: 0-9728053-8-9 Price: $14.99
Author: Lawrence Harte #Pages: 117 Copyright Year: 2004

This book explains why people and companies are using SIP equipment and software to efficiently upgrade existing telephone systems, develop their own advanced communications services, and to more easily integrate telephone network with company information systems. This book provides descriptions of the function parts of SIP systems along with the fundamentals of how SIP systems operate. It identifies and explains what key services are possible through the use of SIP and how existing phone systems can be upgraded to SIP capabilities. It describes why it is easy to integrate SIP with information systems along with how to develop new advanced revenue producing services. The reader will learn the basic SIP system development and installation process and how to manage SIP systems. Explained are the typical costs of SIP systems and how SIP technology is changing to meet future multimedia communication needs.

Telecom Systems

ISBN: 0-9728053-9-7 Price: $34.99
Author: Lawrence Harte #Pages: 480 Copyright Year: 2004

This book Telecom Systems shows the latest telecommunications technologies are converting traditional telephone and computer networks into cost competitive integrated digital systems with yet undiscovered applications. These systems are continuing to emerge and become more complex.Telecom Systems explains how various telecommunications systems and services work and how they are evolving to meet the needs of bandwidth and the applications involved for the hungry consumers. Literally thousands of people who need to understand how telecommunications systems operate and the services that they offer have found previous works of these authors to be extremely valuable. Telecom Systems provides details of the latest telecommunications technologies.

Introduction to Transmission Systems

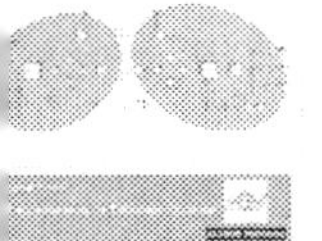

ISBN: 0-9742787-0-X Price: $14.99
Author: Lawrence Harte #Pages: 52 Copyright Year: 2004

This book explains the fundamentals of transmission lines and how radio waves, electrical circuits, and optical signals transfer information through a communication medium or channel on carrier signals. It also explains the ways that a single line can be divided into multiple channels and how signals are carried over transmission lines in analog or digital form. Find out how the transmission of information can be distorted by noise and distortion impairments and how error detection and correction techniques used in digital transmission can improve the performance of transmission lines. You will learn how some of the energy of signals transferred through transmission lines may leak out and be transferred to other lines causing crosstalk.

Tehrani's IP Telephony Dictionary

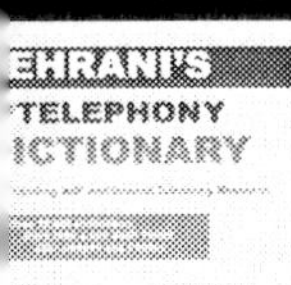

ISBN: 0-9742787-1-8 Price: $39.99
Author: Althos #Pages: 628 Copyright Year: 2003

Tehrani's IP Telephony Dictionary, The Leading VoIP and Internet Telephony Resource provides over 10,000 of the latest IP Telephony terms and more than 400 illustrations to define and explain latest voice over data network (VoIP) technologies and services. It provides the references needed to communicate with others in the communication industry where many new terms and concepts are being added each day. It includes directories of magazines, associations and other essential trade resources to help industry professionals find the information they need to succeed in this rapidly growing industry.

Practical Patent Strategies Used by Successful Companies

ISBN: 0-9746943-3-9 **Price:** $14.99
Author: Eric Stasik **#Pages:** **Copyright Year:** 2004

This book explains how companies can use patent strategies to achieve their business goals. Patent strategies may considered abstract legal or economic concepts. Examining how patents are used by leading companies in specific bu ness applications can provide great insight to their practical use and application in your business plan. This book pre ents in plain and clear language why having a patent strategy is important and how to implement programs and proce es to develop and manage different forms of intellectual property (IP). The reader will learn why a patent strategy essential to most business plans, how a patent strategy can affect the design and development process, methods "design-in" patent protection, how to reduce, or avoid, royalty payments, and ways to stay one step ahead of the co petition.

Introduction to xHTML

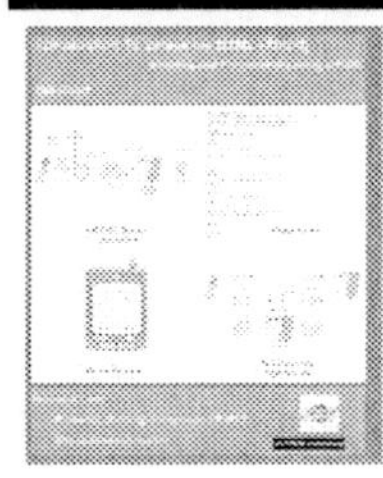

ISBN: 0-9328130-0-4 **Price:** $34.99
Author: Lawrence Harte **#Pages:** **Copyright Year:** 2004

This book explains what is xHTML Basic, when to use it, and why it is important to learn. You will discover how the xHT Basic language was developed and the types of applications that benefit from xHTML Basic programs. The basic p gramming structure of xHTML Basic is described along with the basic commands including links, images, and spec symbols that are used. An introduction to scripting is included and how to create advanced services and features usi xHTML Basic programming. You will learn the similarities and difference of xHTML Basic to other communication la guages including WML and HTML. You will also learn how to write and simulate xHTML Basic programs. Explained is t operation of WAP servers and how to publish xHTML Basic programs to servers to allow them to be accessed

Introduction to SS7

ISBN: 1-9328130-2-0 **Price:** $14.99
Author: Lawrence Harte **#Pages:** **Copyright Year:** 2004

This introductory book explains the basic operation of the signaling system 7 (SS7). SS7 is the standard communicat system that is used to control public telephone networks. This book will help the reader gain an understanding of S technology, network equipment, and overall operation. It covers the reasons why SS7 exists and is necessary, as well step-by-step procedures that describe the actions that occur in the network. This introductory book explains how the p lic telephone network is coordinated through the use of signaling messages. The SS7 common channel signaling h been standardized since the 1970's and is used by almost all public telephone companies throughout the world. If y are involved in telecommunications systems, SS7 is part of your system.

Creating RFPs for IP Telephony Communications Systems

ISBN: 1-9328131-1-X **Price:** $19.99
Author: Lawrence Harte **#Pages:** **Copyright Year:** 2004

This book explains the typical objectives and processes that are involved in the creation and response to request for p posals (RFPs) for IP Telephony systems and services. It covers the key objectives for the RFP process, whose involv in the creation and management of the RFP, and how vendors are invited, evaluated, and notified of the RFP venc selection result. You will learn what are RFPs and RFQs and why and when companies use and RFPs for IP Telepho Systems. Covered are the key objectives that RFP must satisfy along with the general creation processes used by m companies to create and manage the RFP process. Companies usually involve multiple departments in the creation the RFP process to identify communication requirements for the entire company. You will discover who is involved in Creation of an RFP and the typical steps performed during the creation of the RFP document.

ATM Basics

ISBN: 1-9328131-3-6 **Price:** $29.99
Author: Lawrence Harte **#Pages:** **Copyright Year:** 2004

Asynchronous Transfer Mode (ATM) is a high-speed packet switching network technology industry standard. ATM n works have been deployed because they offer the ability to transport voice, data, and video signals over a single syste The flexibility that ATM offers incorporates both circuit and packet switching techniques into one technology, creatin variable transport network solution with simple network processing functions. ATM Basics provides an understanding how the systems operate and the applications that use ATM systems. ATM has become the world standard that is us to interconnect telephone and data networks. This book covers the operation of ATM systems and it explains why A switching is much faster than other types of packet data networks such as TCP/IP.

wireless Markup Language (WML)

ISBN: 0-9742787-5-0 Price: $34.99
Author: Bill Routt #Pages: 292 Copyright Year: 2004

Wireless Markup Language (WML) Scripting, Scripting and Programming using WML, cHTML, and xHTML explains the necessary programming that allows web pages and other Internet information to display and be controlled by mobile telephones and PDAs. Althos Publishing announces the addition of its newest book in its full line of communication books, Wireless Markup Language (WML) Scripting and Programming using WML, cHTML, and xHTML. Wireless markup language (WML) and WMLScript are programming languages that are used to provide information services to portable wireless devices. This book explains how and why companies use WML to develop and provide information services to mobile communication devices. WML protocols and scripts are used to create web pages that are compatible with them.

Introduction to Bluetooth

ISBN: 0-9746943-5-5 Price: $14.99
Author: Lawrence Harte #Pages: 60 Copyright Year: 2004

Introduction to Bluetooth explains what is Bluetooth technology and why it is important for so many types of consumer electronics devices. Since it was first officially standardized in 1999, the Bluetooth market has grown to more than 35 million devices per year. You will find out how Bluetooth devices can automatically locate nearby Bluetooth devices, authenticates them, discover their capabilities, and the process used to setup connections with them. You will learn how the use of standard profiles allows Bluetooth devices from different manufacturers to communicate with each other and work together in the same way. This book explains how Bluetooth's spread spectrum technology allows Bluetooth devices to share the frequency band with wireless LANs, microwave ovens, cordless telephones, and other devices.

Introduction to CDMA

ISBN: 1-9328130-5-5 Price: $14.99
Author: Lawrence Harte #Pages: 52 Copyright Year: 2004

Introduction to CDMA book explains the basic technical components and operation of CDMA IS-95 and CDMA2000 systems and technologies. You will learn the physical radio channel structures of the CDMA systems along with the basic frame and slot structures. Described are the logical channels and their functions along with the key CDMA network components and how they communicate with each other. You will learn how CDMA systems continue to evolve to provide for new types of messaging, high-speed data, and new multimedia information services.

Introduction to Mobile Telephone

ISBN: 0-9746943-2-0 Price: $10.99
Author: Lawrence Harte #Pages: 48 Copyright Year: 2004

Introduction to Mobile Telephone explains the different types of mobile telephone technologies and systems from 1st generation analog to 3rd generation digital broadband. It describes the basics of how they operate, the different types of wireless voice, data and information services, key commercial systems, and typical revenues/costs of these services. Mobile telephone technologies, systems, and services have dramatically changed over the past 2 years. New technology capabilities and limited restrictions (deregulation) are allowing existing systems to offer new services. Many of these new services compete with other types of wireless systems that have not experienced significant competition. While new competition has provided lower cost services for consumers, it means a rapidly changing marketplace for the wireless industry. Some of these changes include the increase in mobile telephone customers from 300 million to 1.3 billion customers within 5 years

Introduction to Wireless Billing

ISBN: 0-9746943-8-X Price: $14.99
Author: Avi Ofrane, Lawrence Harte #Pages: 48 Copyright Year: 2004

Introduction to Wireless Billing explains billing system operation for wireless systems, how these billing systems are a bit different than traditional billing systems, and how these systems are changing to permit billing of non-traditional products and services. This book explains how companies bill for wireless voice, data, and information services. Billing and customer care for wireless systems convert the measured amounts of services (bytes of digital information transmitted or information services provided) within a network into the money that will be received by the service provider. The billing systems used in wireless systems can vary from simple one time charges for hot-spot wireless access points to integrated mobile cellular systems that allow real-time (near instant) activation and prepayment of services.

ORDER FORM

Phone: 919-557-2260
800-227-9681
Fax: 919-557-2261
404 Wake Chapel Rd., Fuquay-Varina, NC 27526 USA
Email: success@Althos.com web: www.ALTHOS.com

Date:_______________

Name:______________________________

Title:______________________________

Company:______________________________

Shipping Address:______________________________

City:____________________ State:__________ Zip:__________

Billing Address:______________________________

City:____________________ State:__________ Zip __________

Telephone:____________________ Fax:____________________

Email: ______________________________

Purchase Order #_______________ (New accounts: please call for approval)

Payment (select): VISA ___ AMEX ___ MC ___ Check ____

Credit Card #: ____________________ Expiration Date: __________

Exact Name on Card: ______________________________

Qty.	BOOK #	ISBN #	TITLE	PRICE EA	TOTAL
Book Total:					
Discounts:					
Sales Tax (North Carolina Residents please add 7% sales tax)					
Shipping: Please apply accurate shipping rates and surcharge per client and order.					
Total order:					

Providing Expert Information

Printed in the United Kingdom
by Lightning Source UK Ltd.
117933UK00001B/109

9 781932 813128